SPACE SCIENCE, EXPLORATION AND POLICIES

TIME-OUT OF TIME

A POSTSCRIPT TO NUCLEAR TIME TRAVEL

SPACE SCIENCE, EXPLORATION AND POLICIES

Additional books and e-books in this series can be found on Nova's website under the Series tab.

SPACE SCIENCE, EXPLORATION AND POLICIES

TIME-OUT OF TIME

A POSTSCRIPT TO NUCLEAR TIME TRAVEL

BERND SCHMEIKAL

DOI: https://doi.org/10.52305/NGTK6470

Library of Congress Cataloging-in-Publication Data

ISBN: 978-1-68507-320-6

Published by Nova Science Publishers, Inc. † New York

Contents

PRELIMINARY REMARK

This publication is a postscript, a follow-up, and addendum to my work *Nuclear Time Travel and the Alien Mind* [1]. And at once it is an examination of the coronavirus crisis, especially under the aspect of Pareto inequality, disaggregation of society and space activities.

PROLOGUE

It is New Year's Eve 2019. Peter Turrini wrote me a letter. The last lines read: "I never know where you are right now. According to the season, it is quite possible that you have withdrawn into the warm climes of Greece, but it could also be that your unsteady nature does not come to rest there and you can live out your unrest in and around Floridsdorf. I simply wish, noble Bernardus, that in this increasingly bad world you will also be well at times in between." In fact, the world seems to be getting worse. Our social world is falling apart.

In May 2020, Donald Trump presented Operation Warp Speed (OWS), a public-private partnership initiated to facilitate and accelerate the development, manufacturing, and distribution of COVID-19 vaccines, therapeutics, and diagnostics. It is no coincidence that the medical activity here is associated with superluminal spacecraft propulsion. There is a plan.

We listen to the radio: Workplaces, companies, and corporations are being downsized or closed. An army of unemployed and low earners is faced with a shrunken labor market. The gentlemen above can choose the best they need. There is no more space for the small ones down there; there is even less space than ever before. Online trade profits from a second coronavirus wave. While many small stationary retailers are suffering from the consequences of the coronavirus pandemic, a record year is looming for online trading. At the top of the list is Amazon – the company posted a profit of the equivalent of 5.3 billion Euros between July and September 2020. This is more than all the large retailers who are making particularly high profits this year. For small stationary retailers, on the other hand, it is far more difficult to make a profit in online trading. This is what I denote as 2nd order Pareto de-equalizing.

Pensions are being questioned and are already being attacked. In the already highly stratified class society, further selection is being made. All scissors of social inequality are wide open. It is not a pleasant situation. The global COVID-19 epidemic offers the ideal means for renewing class society and forms of rule for unimagined and previously unknown elites. Conflicts escalate everywhere, from small neighborhood disputes to intercontinental acts of war. Mainstream media try to calm down their listeners and readers and restore custom and decency. For the first time in a long time, the subject of shame and embarrassment is again being discussed by sociologists. People are worried. Secret services are trying to expand their spheres of power. Politicians are being poisoned. New biological warfare agents are being developed for the military. It is downright naive to believe that civil society is not affected. Military and civil laboratories are engaged in research into effective virus warfare. It is naive to think that our governments only want to do us good. Rather, they are ready to submit to the greatest of existing powers, even if their activities turn out to be extremely destructive. One of the worst conspiracies concerns the conspiracy against conspiracy theories and theorists. No conspiracy theory is needed at all to recognize that viruses are used not only in wartime but also in peacetime against the civilian population – it is not absurd to believe that the COVID-19 virus resulted from experiments with the SARS forms and escaped from some laboratory.

In the midst of this completely confused situation, in which all orientation is lost, all possible forms of escape reactions arise according to an old animal pattern. The most important and probably now the most prominent form of escape is the escape into space. This escape, of course, is veiled and presented with scientific means as a predominantly sensible, rational undertaking. Within the current pattern of social development, the disintegration of society into several strands can no longer be stopped. Part of the people will use the means of time travel technology for their journey into space. Such travel will also carry the stigmata of good and bad with it. This may seem pessimistic to us at first, but, overall, it will open up new paths in cosmic life. We ponder "the world." But what is world?

The world could be the entirety of all accesses in living presence made by all living beings, who focus their attention on anything moving within it. World is an excellent amount of living relations, created by the intelligent energy of awareness. I want to say that the world is self-opening in living awareness. This opening may be accessible to individual beings in rare exceptional cases when there is love.

Chapter 1

LIGHT OFF-TIME

I ask readers to take their eyes off the box. Isn't it noticeable that there are some massive lines of demarcation our thinking cannot avoid? There is physics, which declares itself responsible for time and space, on the other hand, there are life sciences, social sciences, and psychology, which may tell us as much, if not more about time, even more than physics. So that is now one of the boundaries that thought is drawing, namely the line between natural and social sciences. But there are also completely different limits, limits that are directly connected with time. Isn't it interesting that at an advanced age we draw a very clear line between ourselves, our own age group, and the young people? In old age we feel the proximity of death, and we are glad that the young, the adolescent, and the young children are carefree; they do not seem to have our idea about age and death. So, let's draw the line between life and death. Who draws this line? Thought creates that limit. Life on its own does not draw such a line.

Why doesn't the writer draw a line between life and death? Why doesn't he make that difference? Has he maybe exceeded that limit? What happens when we get up in the morning? The bed is still warm, we pull back the duvet, hop out of bed, get rid of the night gown, go to the bathroom, brush our teeth, wash our face, and so on. We start to think about the day, make plans, decide what exactly we will do, and start hoping that everything we do now can really be implemented. After a while we may leave the house, get into our car, start the engine, drive around the corner, turn onto the main street, drive forward. We have a long way to go. Maybe we are on our way to work. Do you notice that our thinking revolves around these events all the time? It is tied to that process. We believe we should drive forward and move forward in time. We believe that some of the matter that surrounds us, that we are curiously involved with, moves with us, forward in space and time. Presumably all molecules, all atoms, all nucleons, all protons, all electrons, move with us forward in space and time. But on closer inspection that is an illusion. There are fields that overlap and these constitute the vehicle in which we sit. Some of them come from afar. They come from spaces and times not accessible to our observations. Do you understand what we're thinking about here? The ways of matter are the ways of nature. These have no measure. These paths run differently in space and in time, differently than thought can imagine. Our thinking has designed its own restricted picture of movement, and basically everything just moves within that picture. Reality – and by that, I mean the physical, the biological, the living, the social, the human reality – has nothing to do with this constructed physical movement. Do you see the line between culture being thought and nature?

There is a boundary between the real world and the imagined world. There is a boundary between an actual spatial and temporal course of events and an imaginary one. Some of us have got used to assigning this course of events to metaphysics. But if you cross the line I am talking about, you don't separate metaphysics from physics. That is the whole secret.

The world is not an idea. It is not even a metaphysical idea. It has little to do with imagination. The moment we realize that thought, thinking about its own processes, is bound to consciousness and creates a picture of the world in its circular motion, we also realize that thought constitutes time. The real world moves beyond this dimension of time. The real events are vividly moved. But this movement is not tied to the time arrow and has nothing to do with the construction of the cognitive system. What is called time creates its own world.

Every living being has its own perception of the world. Each of us has his or her own ideas. People see the world with one or two eyes. But a spider may have eight eyes and a panoramic view. Our senses create a worldview, our immediate sensual worldview. This immediate living perception arises in the presence of intelligent awareness. As perception takes place, we think about it, and we create our own cognitive images. These images appear to be separate from one another. But we talk about them. This creates a common picture of a unified, space-like, and temporal world in which, so we believe, we are moving. In our view, the world becomes an object, something objectively real. But this is an illusion.

Let us ask ourselves whether the spider with its eight eyes and its observer, the writer, live in the same world. We honestly can't say anything about that. The objective, the supposedly real world does not exist. It is something metaphysical, something that we construct in our thoughts. Now, it actually seems that the world is getting more and more complicated because so many people take so many different pictures of it. As a cognitive construct, it disaggregates, it becomes more and more complicated.

Isn't it interesting that the universe, which supposedly also exists for every individual, is constantly expanding, and this expansion is accelerating? What's going on there? Is there perhaps a connection between the physical facts, facts of cosmology, and the cognitive facts, both of which are becoming more and more complicated? In *Nuclear Time Travel* [1], I mentioned that Albert Einstein made mistakes, the meaning of which we still cannot really estimate. There is an interesting man, Ben Rich, who was the director of Skunk Works at Lockheed Martin. He had also a contract with the CIA and was an engineer at the USAF. There is a famous quote of what he was supposed to have said on his deathbed when someone asked him what the technology was that would allow us to travel to the stars.

To this, he said: "There is a mistake in the equations and we have found it and now it is possible for us to travel to the stars and it will not take a lifetime to make such a journey." In *Einstein's Blunder* [2], I use algebraic means to show why relativity, be it special or general, is incomplete for algebraic and phenomenological reasons. There are algebraic works in which I have shown why there are cross-time interactions that contradict our traditional ideas about causality. Let's go a little further. We believe that there is a space between two objects. The two objects are said to be spatially separated. There is a space-time difference, a distance. Let us consider two elementary particles that are "in space." But things are not that easy. We also speak of "empty space." But it is not at all clear what an empty space is. That is why in reality the space is completely defined by material conditions and interactions. Empty space is a mental construction. It may be that the elementary particles, together with the spider with eight eyes and the writer, are all in the same space; but, in reality, there is no space, because the space is a mental construction. Orientation connects the particles, me, and the spider. All three have a concept and a perception of what is in front and what is behind, whether there is something above or to the left and right of us. We are talking about orientation morphemes. Structures of experience operate within us, they work in an inner space, the brain's mind, and they connect with the outside world. Both worlds are material, are real and oriented. The orientation morphemes mediate between inside and outside. Thus, the metaphysical world, which appears to exist objectively, becomes binding for all of us. Elementary particles, spiders, writers, animals, and plants, even space and time, appear to us as oriented entities.

Let's go back to the question of interstellar travel. Ben Rich was, of course, a technically and scientifically extremely competent person, a design engineer. Nevertheless, the question of whether there were UFOs was out of the question for him. He knew UFOs and he knew extraterrestrial humanoids. There are people who say that they were on board flying saucers and similar aircraft and communicated with extraterrestrials. Interestingly, most of these witnesses made statements about space, time, light, and life phenomena above all. For example, Orfeo Angelucci [3, p. 11]: "Interplanetary ships and saucers of various material densities can approximate the speed of light. This seems impossible to you only because of a natural principle which has not yet been discovered by your scientists. Also, the Speed of Light is the Speed of Truth. This statement is presently unintelligible to Earth's peoples, but is a basic cosmic axiom. Approaching the speed of light, the Time dimension, as known upon Earth, becomes non-existent; hence in this comparatively new dimension there are incredibly rapid means of space travel which are beyond man's comprehension. Also, within the Records of Light are to be found a complete history of Earth and of every entity which has incarnated upon it."

Think for a moment about Einstein's famous formula:

$$E = mc^2 \tag{1}$$

It is actually an approximation. It is mathematically bad, and it does not correspond with the whole phenomenology of light events. We are so used to that formula that we hardly ever think of considering the inverse operation:

$$c = \sqrt{\frac{E}{m}} \tag{2}$$

In this equation we can take the mass m as the product of a density of matter with a volume, as follows:

$$m = \rho . V \tag{3}$$

It is immediately apparent that the speed of light diverges when the mass-density ρ approaches zero:

$$c = \sqrt{\frac{E}{\rho\, V}} \tag{4}$$

That looks comparatively harmless, but it actually seems to have enormous consequences. The formula is at first understood as a metaphorical reference. The associated phenomenology is rather complicated. We seem to be dealing with very complex non-linear processes, whereby the energy areas, which are to be regarded as interacting, i.e., interacting fields, cannot easily be located in the space cognition has prepared. As we have designed space so far, the concept is not sufficient to understand the global and local events of quantum theory. We should talk a lot about emptiness. The interactions between the different areas are described by a phenomenology that we do not fully understand. It really is a completely new physics. This doesn't make Einstein's reasoning wrong. His two theories of relativity – the special and the general theory of relativity – remain partially binding. But similar to the way in which Newton's theory of gravity was a guide in Einstein's considerations, Einstein's views on the phenomena of light propagation and the measurement of space and time with the help of light signals are now also only a guide for our new ideas. We have to think of the universe as even much more surprising than a biological process. The light is, in a way, an extremely sensitive phenomenon. We know from laboratory experiments that the speed of light can actually take on all possible values between 0 and c locally on Earth. We can easily stop photons. We can stop the so-called group speed of a wave packet in a very special containment in the laboratory. The photon is then able to stand still and, interestingly, it has a resting mass.

Let's look at Orfeo Angelucci's sentence again. It is particularly interesting:

"Interplanetary ships and saucers of various material densities can approximate the speed of light. This seems impossible to you only because of a natural principle which has not yet been discovered by your scientists. So,

the Speed of Light is the Speed of Truth.

This statement is presently unintelligible to Earth's peoples, but is a basic cosmic axiom."

That statement is quite understandable. Since 1996 and afterwards, I have shown what the speed of light has to do with some peculiar binary logic-generating systems and how, via a parallelism of the algebraic space-time and logic, a dependence of the truth-classifiers on the conditions of the theory of relativity, in particular the speed of light measured in metric space, comes into being. The algebraic structure of space-time is connected with logic. In this context it can systematically be shown how generating polarity strands, i.e., fermions, are linked with generating units of Minkowski algebra. We have to imagine that there are mathematically simple, yet global, non-localized entities to which polarities can be assigned. But at first nothing will be known about their localization, nor will we need to know anything about it. In any case, these fundamental quantities, which we use and which are initially only of a purely mathematical nature, have a binary numerical form, with the help of which both intelligent processes and measurable spatiotemporal regions can be generated in an algebraic way. In nature, this denotes a parallelism between intelligent energy and quantum motion.

We usually assign events to some specific points in space-time. In the so-called inertial systems that are not accelerated that is to say, they are "free of force," we then assign a time coordinate $c\,t$ and a triple of Cartesian coordinates x, y, z to any event to be localized. Practically, in space-time algebra, the position of a "mass point" is given by a four-vector of the following simple form:

$$\vec{x} = x_1 e_1 + x_2 e_2 + x_3 e_3 + c\,t e_4 \tag{5}$$

We therefore assume that we can align a space-time framework locally; in short, we have a given direction, say e_1, and e_2, e_3 are orthogonal to it. The three-unit vectors have a certain representation, usually in the form of matrices, and a positive signature. That is, roughly speaking, a line element whose square is equal to the identity Id of the

Clifford algebra. However, the quantity e_4 belonging to the time measure has a negative signature, that is:

$$e_4^2 = -Id \tag{6}$$

In accordance with the fact that the generating basis of such a Minkowski space $\mathbb{R}^{3,1}$ has signature $\{+ + + -\}$, mathematicians say that such space has an indefinite metric. The vector $\vec{x}$ obtains a quadratic norm, namely the square of its length in this metric:

$$|\vec{x}|^2 = x_1^2 + x_2^2 + x_3^2 - c^2t^2 \tag{7}$$

It is therefore that we are confronted with a hyperbolic geometry rather than with the elliptic geometry introduced by Euclidean definite metric. Now, the interesting thing about the Clifford algebra of Minkowski space-time is that it also means that oriented surfaces and oriented space-time volumes have a significant meaning. This also gives the Dirac equation a more understandable form that can now be interpreted much better. In the geometric algebra, briefly referred to as Minkowski algebra, we have to take into account all 16 external products of the unit vectors generated by the original basis. These are:

$$\{Id, e_1, e_2, e_3, e_4, e_{12}, e_{13}, e_{14}, e_{23}, e_{24}, e_{34}, e_{123}, e_{124}, e_{134}, e_{234}, e_{1234},\}$$
basis of $Cl_{3,1}$

(8)

Many publications, such as Lounesto [4] or my book *Decay of Motion* [5, p. 70], explain how these 16 basic units of Minkowski algebra can be represented using real-valued matrices so that all necessary requirements of geometric algebra are met. It is best to use the representation in Schmeikal [6], because it can easily be used to build both logic and space-time. Now, it is fairly obvious that the base units of grade 1 do not commute with each other, since this fundamental non-commutativity is what makes spin phenomena. So, the magnitudes e_1, e_2, e_3, e_4 do not commute, of course, but anti-commute.

Now, it is interesting to note, and I have spent several years with this, that fermions from six fundamental groups can be built in a very simple way from four basic units that commute with one another; and that is extremely desirable and is to be expected. Because these four quantities are to constitute a single fermion, it is also to be expected that they can be assigned to the same observation level so that they can be observed simultaneously. In fact, each is a set of quadruples, a particularly simple type of quaternion, for each of the six quark states. I will now write down one of these six so-called color spaces so that you can play with them:

$$ch_1 = \{Id, e_1, e_{24}, e_{124}\} \text{ . . . the "first" color space} \quad (9)$$

It turns out that these four oriented unit multivectors of degrees 0, 1, 2, and 3 commute and do not need the large 16-element matrices of the Majorana representation; on the contrary, a representation of four real numbers in the fourfold number ring ${}^4\mathbb{R} = \mathbb{R}\oplus\mathbb{R}\oplus\mathbb{R}\oplus\mathbb{R}$ is sufficient. It is a thin slice of the whole representation, so to speak. It is possible to represent the four commuting base units as follows:

Correspondence 1.

Geom. unit	Id	e_1	e_{24}	e_{124}
vector in ${}^4\mathbb{R}$	$(+1,+1,+1,+1)$	$(+1,+1,-1,-1)$	$(+1,-1,+1,-1)$	$(+1,-1,-1,+1)$

(10)

The superposition of vibrations in these four interchanging space-time areas then constitutes the elementary energy densities of the fermions of the u-quark, i.e., a lepton together with the three colors of a u-quark.

$$f_{11} \overset{\text{def}}{=} \tfrac{1}{2}(Id + e_1)\tfrac{1}{2}(Id + e_{24}) \qquad f_{12} \overset{\text{def}}{=} \tfrac{1}{2}(Id + e_1)\tfrac{1}{2}(Id - e_{24})$$
$$f_{13} \overset{\text{def}}{=} \tfrac{1}{2}(Id - e_1)\tfrac{1}{2}(Id - e_{24}) \qquad f_{14} \overset{\text{def}}{=} \tfrac{1}{2}(Id - e_1)\tfrac{1}{2}(Id + e_{24}) \tag{11}$$

Let's try that. It follows:

$$f_{11} = \frac{1}{4}[(+1,+1,+1,+1) + (+1,+1,-1,-1)][(+1,+1,+1,+1) + (+1,-1,+1,-1)] = = \frac{1}{4}(2,2,0,0)(2,0,2,0) = (1,0,0,0) \tag{12}$$

These four sizes build up the famous Kotessarines, the simplest of the eight possible shapes of quadruple rings of numbers [7], or, sloppily speaking, one kind of quaternions. We get the four binary, primitive idempotents of Clifford algebra $Cl_{3,1}$.

Correspondence 2.

prim. idempotents	(1,0,0,0)	(0,1,0,0)	(0,0,1,0)	(0,0,0,1)
Fermions	lepton	u red	u green	u blue

(13)

See how simple and wonderful that design is; the geometric base units are all a combination of polarized entities, as if they were charged or carried some feature of isospin, which they actually do. These creatures are nonlocal and capable of attracting or repulsing each other and thereby constitute what we observe as partons within the nucleons under observation. And the fermions themselves are polarized in a different way. They carry energy, having a unity 1 in some character. But no energy, i.e., zero, combines with the other three characters. Now think about those two quantities e_{24} and e_{124} that involve the time direction e_4 with its measure ct. Physicists used to set $c = 1$, hence carry out a scaling on the base unit. So, they do the rigor with a simplified square norm, namely:

$$|\vec{x}|^2 = x_1^2 + x_2^2 + x_3^2 - t^2 \tag{14}$$

and measure time by light-seconds. Now forget about the equations and let us get out into space. And it couldn't be easier. Of course, there are also very complicated 10-dimensional representations and more. I don't want to talk about that here.

We are so sure that we know exactly how far away a star is from the sun that we claim to be, say, 10,000 light-years away. This is utter nonsense. We don't even know how far a star is from the sun, which we say is only 40 light-years away. Because the matter is much more complicated than we think. Please look again at the equation for light-speed. It can also be written in the following form:

$$c = \sqrt{\frac{\rho_E}{\rho_m}} \tag{15}$$

that is, the speed of light is the square root of the energy density divided by the density of matter. Now it is quite the case that in the so-called empty space, in this nirvana between the stars, the energy density is by no means zero. But ρ_m, the density of matter, is. This means that the speed of light out there is much greater than we measure it in our local empty room in the laboratory. In addition, there are other things that are very mysterious, namely that the energetic structure of the universe is more like an organism than this dead rigid thing, space-time, that we used to consider binding at the time. What actually happens when, for example, the speed of light, which we normalized to 1 in the last equations, changes, when it drops to $c/2$. Let us see.

Correspondence 3.

Geom. unit	Id	e_1	e_{24}	e_{124}
vector in $^4\mathbb{R}$	$(+1,+1,+1,+1)$	$(+1,+1,-1,-1)$	$\left(+\frac{1}{2},-\frac{1}{2},+\frac{1}{2},-\frac{1}{2}\right)$	$\left(+\frac{1}{2},-\frac{1}{2},-\frac{1}{2},+\frac{1}{2}\right)$

(16)

What would happen to the red u-quark $f_{12} \stackrel{\text{def}}{=} \frac{1}{2}(Id + e_1)\frac{1}{2}(Id - e_{24})$? With $c = 1$ we had that $|u_{red}\rangle$ =(0,1,0,0). But with $c \longrightarrow c/2$ we obtain:

$$f_{12} = \frac{1}{4}[(+1,+1,+1,+1) + (+1,+1,-1,-1)][(+1,+1,+1,+1) - (+\tfrac{1}{2}, -\tfrac{1}{2}, +\tfrac{1}{2}, -\tfrac{1}{2})] =$$
$$= \frac{1}{4}(2,2,0,0)\left(\tfrac{1}{2}, \tfrac{3}{2}, \tfrac{1}{2}, \tfrac{3}{2}\right) = \left(\tfrac{1}{4}, \tfrac{3}{4}, 0,0\right) \qquad (17)$$

So, we observe a transmutation $(0,1,0,0) \longrightarrow \left(\frac{1}{4}, \frac{3}{4}, 0,0\right)$ that can be interpreted as the inner of a nucleon tending to have three times more green than red u-quarks, and this has to do with probabilities and truth values of the logic involved.[1]

I want to at least give you a little hint of how these sorceries work. I used these few short series of unitary numbers, which, on the one hand, have a binary nature, but, on the other hand, actually already denote polarity strings with which we can represent fermions in quantum physics. These short series of numbers are also logical truth tables, and, in addition, they represent some of the most important Clifford numbers of relativistic geometry. Let us consider the four generating binary connectives that I have shown in Table 11 of the paper *Four Forms Make a Universe* [6]. They are the generating binary connectives in a geometric

[1]This is a wonderful thing because it turns out that these entities, which occur in binary polarization, interact with each other in an intelligent, logical way such that the natural truth values that emerge in the process actually depend on the speed of light. Some of the fundamental considerations appeared in the article "Transpositions in Clifford Algebra" in the book *Clifford Algebras – Applications to Mathematics Physics and Engineering*, and later in the book *Decay of Motion – The Anti-Physics of Space-Time*. On pages 108–109 you will find how those polarity-strings can be derived strictly and what they have to do with the color spaces of the strong force. The forces of nature therefore express themselves structurally and mathematically in the interactions of some basic polarity strands. I say all these things here quite explicitly so that in the future some talented physicists will be able to fall back on the right works of mine. You will then surely recognize how neural networks can develop in what today we call space-time, whereby the interacting entities are represented by fields whose topological properties we initially know little about. But, in a nutshell, I can suggest that the new topos has to do with entanglement, only among other things. In other words, two photons that are entangled with each other, or even two electrons, atoms, molecules, and so forth, that are entangled with one another, are basically no distance apart. We have to assume that. Based on a new concept of topological proximity, we can work out a new concept of space.

algebra, namely Minkowski algebra. Roughly speaking, this is the geometry of the special theory of relativity. These generators denote:

1. the true statement,
2. a statement A,
3. a statement B, and
4. the logical identity, "A is logically identical to B."

The fourth is denoted as the XNOR operation. So, we have

$$Id \quad A \qquad B \qquad A \equiv B \tag{18}$$

These four statements can nicely be represented by our familiar short series of numbers. These are the quadruples:

Correspondence 4.

connectives	Id	A	B	$A \equiv B$
Kotessarines	$[+1, +1, +1, +1]$	$[+1, +1, -1, -1]$	$[+1, -1, -1, +1]$	$[+1, -1, +1, -1]$
	Id	e_1	e_{24}	e_{124}

(19)

These quaternionic binary connectives are basic generators for any system of logic, whether Boolean, Hegelian, or any other logic system. Now you see the connection with space-time. Clearly, the truth classifiers of B and the XNOR having B input, change when the velocity of light is altered. This is, in pure mathematical terms, Orfeo Angelucci's sentence "the Speed of Light is the Speed of Truth." For it is light that calibrates the truth-classifiers of any material logic system. Only a fixed light-velocity allows for fixed truth values. But light is flexible and accesses all possible values of acceleration and deceleration.

See the simplicity and the beauty of the mathematics outlined. We have two fundamental operations, one of which is geometric, while the other is logic. Clearly, the wedge product of e_1 with e_{24} is e_{124}.

In this case it can be affirmed by component-wise multiplication of the corresponding entries, that is,

$$[+1,+1,-1,-1]\,[+1,-1,-1,+1]=[+1,-1,+1,-1].$$

On the other hand, the logic XNOR operation acting on binary the ground-field $\{\pm 1\}$ confronts $[+1,+1,-1,-1]$ with $[+1,-1,-1,+1]$ and thus brings forth $[+1,-1,+1,-1]$, which means both $A \equiv B$ and the oriented space-time-volume e_{124}.

Let us remember Orfeo Angelucci's statement in *The Secret of the Saucers* (1955) [3]:

> "Approaching the speed of light, the Time dimension, as known upon Earth, becomes non-existent; hence in this comparatively new dimension
> there are incredibly rapid means of space travel
> which are beyond man's comprehension. Also,
> within the Records of Light are to be found a complete history of Earth and of every entity which has incarnated upon it."

There are other contacts who have said something about space travel by extraterrestrials, for example Martin Wiesengrün (=Andreas Heinisch). In his book, *My UFO Experience on Rügen* [8], Martin Wiesengrün describes the experiences he had on the island in 1950.

Wiesengrün had brought on board the UFO a volcanic stone from Earth. With that, some wonderful things happened, which literally led to the secrets of time travel through several "observations and experiments." On pages 104f he writes:

> "In a kind of screw clamp my stone from Earth was ready to be tested. I watched how a fine beam of light precisely cut this hard stone and at the same time polished it to a mirror finish, magnifying it several times for me.
>
> "Half of it was placed on a high-gloss plate and the other half in a bowl made of a special material. It flashed briefly, the stone on the plate had evaporated. Some highly caustic liquid was poured into the stone in

the bowl; gradually a viscous mass developed, which was poured into a square special container. After only a few minutes, a beautiful new, square stone was created, which was placed on a small pedestal. Then invisible forces made my new stone float around in the room; Seconds later it returned to where it started. Suddenly the stone faded as if a kind of thin mist swallowed it up. Oddly, I could see through both media, but gradually everything disappeared. The pedestal was empty, but the stone reappeared on a distant pedestal. …

"Immediately afterwards, Krotk showed me what sound waves can do. …

"Krotk explained another device to me. It worked on the basis of unimaginably small sound waves, combined with an enormous supply of energy. With this device, the matter that the human eye perceives as invisible can be made temporarily visible on a specially designed screen. So you can see the physical nature of the processed matter. This matter, i.e., the air that surrounding it suddenly turns out to be a rather fragile medium. This invisible matter then looks like a faint fog or a cloudy glass pane that has cracked and behind which it simmered weakly. The breaking points or cracks are particularly important. But only those cracks in which there is also unrest.

"And these cracks are important because you are able to enlarge them considerably, which creates a time gap. There you can fly in and choose any place in the past and the future.

"So you can easily visit new and distant worlds. However, if you violate these laws, you are irretrievably missing." [2]

Let's take a closer look at the journey into a time slot. We have to imagine that we are moving in a kind of closed time-sphere. We are inside a spaceship that has its own closed space-time. With this sphere we now navigate into these holes, into these wormholes in space-time. What did I say before? I explained in the previous paragraphs why the energy density of a subnuclear elementary particle has to be applied in 0, 1, 2, and 3 commuting dimensions at the same time. We have said that the energy density must be considered distributed over a scalar component, a vectorial, an oriented space-time area, and an oriented space-time

---

[2] Translated from German to English by the author with help of Google.

volume. This is determined by the blueprint of the energy density. And now let's look at what I happened to find yesterday, namely a book. The book is called *Physics of Quantum Rings* by Vladimir M. Fomin [9]. I came across this book after I happened to see one of these documentaries about the *Mysteries of the Universe* on n-tv-info the other day. Among many other new approaches, so-called time rings were discussed. In a review of *Physics of Quantum Rings* it said: "The editors Vladimir Fomin and co-authors impressively show that the traditional division into 3, 2, 1 or 0-dimensional systems, is not as comprehensive as it seems in their book, now in its second edition. In contrast to the systems mentioned, quantum rings are not simply connected in the mathematical sense. And even the question of whether it is a quantum wire closed to form a ring or a quantum dot from which the inside is cut out shows how diverse and new the physics that quantum rings offer." The new view actually comes about through the topological view of the physical processes. Yes, that's why my mathematical explanations of the previous paragraphs on this point are, on the one hand, very complicated, and on the other hand, very simple for mathematicians. As I said before, I saw a film from the series *Mysteries of the Universe* on television, and I would like to say that the physicists who had their say here told very beautiful stories that were certainly in line with the "narrator" Morgan Freeman. Morgan Freeman is a brilliant, very personable film narrator. Several so-called stars had their say in this broadcast, such as a theoretical physicist from Utrecht, the well-known Harvard particle physicist Lisa Randall, a physicist Maria Spiropulu, and a man who deals with time rings and time fluctuation. They were all wonderful storytellers.

Basically, every story has to turn out to be wrong at some point in time, as Popper has said, simply because the world is moving and the sciences are developing. But a good theory should not be falsified too soon. Time travel is something very special because its phenomenology is literally connected to our neurons. When the concept of time reaches its limits, we humans also experience fundamental existential phase-transitions. We literally become new beings. The newest physicists are trying to reject the old concept of the continuous space-time framework.

Of course, they are right in doing so. But if we only replace space and time by fractal crumbs or with sand in a global ocean, then we are wrong. We make this mistake because chaos theory and similar mathematical models allow us cheap applications. But these are wrong. Time may be unstable, and it is, but a model that we should be building should not be just time crumbs or fluctuations. Time has a different nature.

We cannot deny that we are born in time. However, it does not mean that time has to be moving linearly from birth to death, so that it follows an arrow. There is always big talk about a flow of time, as if the whole world were in this flow of time. And I tell you, not even one single person is in such a flow of time. Rather, the events of what we call the world are shaped by the grasp of the present. This grasp is mysterious, its mechanics unknown. The shape of the present in human consciousness determines what the world appears to be in this consciousness. If we coordinate all of these individual contents of consciousness that we accumulate through the latest films, documentaries, literature, and new media, then a link is established in the present. It is these networks of current experiences that ultimately provide us with a binding worldview that we then agree on. But there is no world as a comprehensive object that would rest reliably in a time stream.

As far as our own "individual" lives are concerned, we have to reckon with the fact that the movements of our present are cyclical in nature. It has loops, the movements of which are extremely complicated, and cannot be fully coordinated with someone else's life, just as we cannot synchronize the pointer of a watch with the water level. The points of contact between our various paths of life are discretely distributed, extremely difficult to grasp, and in no way accessible to the whole general public. The so-called public also has no objective existence. Rather, it is embedded into a complex process of presence. Nevertheless, there appears to be something like entropy and thermodynamics throughout the movement. It is with the help of these concepts, which are thermodynamic in nature, that we can paint a relatively obliging picture of a linearly moving world in time. As I have said many times, the triad "*time, volume and four-volume*" represents such a thermodynamic trinity. But this picture is essentially still incorrect because it is still based on our idea of a stable 4-dimensional space-time. The algebra of such space-time can only give us clues for the new quantum physics. There are physicists who have eliminated time and have investigated the relationship between the phase velocity and group velocity of a photon field. These approaches are extremely important. I will discuss them a little later. What we need here is an understanding of time in spaceships that does not obey the usual and allegedly binding laws of physics. The speed of light is not a top speed. It can be as small as you like and as big as you want, even while it always proves to be at a maximum locally. The "real speed" depends entirely on the current, local structure of matter.

Now, think again that our spaceship and our bodies move into the interior of such a spatiotemporal fiber, a time-slot, and that we, being aware of the previously known top speed of light, override it. Then all of our molecules, all of our nucleons, will participate in this game. However, since the speed of light is much higher than we assumed, the play of forces inside the nucleons will reform, their interaction will reshape. I don't want to go into everything that can happen then. Just keep in mind that the distances we measure with our strange light rays could be much smaller than we previously thought. Let's look at the edge of the universe; let's see 13 billion light-years into the past. Let's look at this edge, which is very cool. We believe that there is a lot of dark matter and dark energy in our universe that change gravitational forces. But it could simply be that the matter structure in these remote regions is such that all distances are much shorter; and it's merely the shortness of these distances that makes the speeds there as they appear to us here and now, all seeming to be much higher than they should be.

Basically, what we are doing is that we keep relativizing the results of recent experiments on our old theories. But we're a little sluggish in reformulating our old theories. We would rather invent new physical relationships, new forms of matter, new energies, rather than admit that the mainstream theories are simply wrong. If we understand that the speed of light is the speed of truth, or let's just say "speed of correctness"

– the speed that also regulates the processes in a quantum processor – then, after a few exact calculations, it becomes clear why time travel is possible. I have been saying since 1996 that the space-time 4-volume cannot be a preserved magnitude, that it cannot be an invariant of motion. It changes in so many ways that it becomes possible to form a capsule that contains its own space-time. Our spaceship embodies its own space-time, inside of which one can live very well. But if you now believe these films that propose representing space-time as semolina, you are somewhat wrong. If you see a film showing an experiment trying to investigate fluctuations in time, you will be faced with the question of whether it might turn out that time begins to fluctuate once we have single atoms or electrons cooled down to absolute zero. This is a very frequently discussed question at the moment. What happens to an ion, an electron, if it is cooled to exactly absolute zero? Does it then stop moving or do completely new properties appear? Is the type of movement preserved? Currently the answer looks something like this: The chemical, fermionic interactions get a completely new and much greater range as soon as the matter is cooled down to absolute zero with deviations of a few nanograde Kelvin. The range of interactions increases. The fluctuations become clearer. I would say there is a quantum phase-transition in which the appearance of the moving energy, its scales, and its dimensions change greatly. That is exactly what the young alien physicist Krotk explained to Martin Wiesengrün.

The limitation of the electronic long-distance effect when exiting the wormhole can be described as a collapse of the wave function "*in these remote regions at the absolute zero temperature*. We then see the local fluctuation to which we have already got used: the Wheeler foam of space-time, something very energetic, the origin of which we do not know. We can call it life. We perceive fluctuations locally, but these fluctuations come from a domain that we simply do not know because it is disconnected and spaced out from our measurements. Even if I say that energy comes from everywhere, it is basically a meaningless statement, although it may be conceivable in certain fairytale images. We simply do not have any measurement specification that would inform us about the origin of free, i.e., unbound, energy. Free energy is released from the context of matter, so if you cool the matter down to the so-called absolute zero, then it condenses to a state of motion of free energy that cannot yet be assigned to the metric space-time. It preexists before any collapse to measurable quanta happens.

It is there where we are confronted with the deeper secrets of space travel. The phenomenology of the creation of energy determines locally where we are, where we go, how fast we can move, how time flies, or whether it does not, how long distances appear to us, and whether we can perceive linear movement at all. The perceptions of matter during such rapid movements outside the solar system are very fragile. They first have to be made and stabilized by the space travelers. Then the brave men have to communicate about it. Then we have to reach an agreement and construct a binding worldview. Only in this picture will new methods of measurement and new protocols be produced.

In Presence

One of the biggest problems with today's protocols is that some formulas of special relativity are roughly applicable, but the universality of the synchronization rule is not easy to understand. Many physicists are rightly puzzled how this should work, that moving matter takes a certain amount of time to cover certain distances, while, on the other hand, unrestrained light can apparently be everywhere at the same time. I have tried to discuss the synchronization rule myself again and again with the help of a few sketches and pictures. The usual picture: someone stands on the earth with a laser pointer and shines it into the sky where the light is reflected down to the floor laboratory, which has meanwhile moved. There it is caught again. Or someone standing in a moving elevator does the same. A mirror is mounted on the ceiling. After a little bit of arithmetic, you get the formulas for shrinking the scale of the ruler and dilating the time. You know the pictures, if you have seen some of the films from space mysteries, then you know the lady who holds a laser pointer vertically in her hand that emits two light beams, one down, one up, and she sits in a train compartment and then the observer sees from the outside how these light rays are bent, since the train is moving very quickly, let's say the upper ray is bent, seen from the outside, by the outer spectator, to the left, and the lower ray is bent to the right in the direction of travel. Or someone is sitting on the train and throws up an orange, and catches it. Observed from the outside, the orange describes a parabola, but within the train compartment it just flies vertically up and down. The passenger in the compartment therefore sees no parabola. The picture turns out to be a relative manifestation. It all looks very convincing, but it has its limits. Because the synchronization process in the field of light is simply much more complex than we can represent in the usual picture of the theory of relativity.

A major obstacle to our understanding is the idea of a current of time, the flow of time in which the events of the world are supposed to be moving. In fact, there is only one presence for the perception-space of a living being, albeit a very complicated one that is difficult to unravel. It is the presence where everything seems to take place. I would like to speak of an intelligent presence here. Even if we say that this or that event took place in the past, the perceptible evidence of these events is all within the presence and possibly perceived as present. For example, we say all observers interested in this event, for example all researchers who are interested in the construction of the Great Pyramid of Cheops, have to collect their perceptions, bring them together, and then compare them with each other, then they will arrive at a binding result that is acceptable to society. It is very similar with space-time. What we say about space-time can ultimately also be found in the present. How does it all work now? Obviously, there is a present part of light, so to say, a complete part of the field that constitutes a connected presence, for which time does not exist at all. In addition, this extremely mobile, very lively part of the field – we would call it a remote field acting at a distance – is responsible for the constitution of the matter we perceive.

We also have to include the so-called gluons, which mediate nuclear forces. Here it is all expressed in a very bumpy way, but I can't communicate it any better. The light that we continuously claim travels at the speed of light has very little to do with the process of synchronizing events. Now, the old contradiction, which literally imposes itself, is well known, according to which the elementary particles move nicely at finite speeds, while the light does not need any time for its movement, so it can be everywhere at the same time. It is clear that something is wrong, and this cannot, in any way, be excused by the naivety of ordinary people or the limited understanding of old-fashioned physicists. It is simply that light that can connect two parts of material that are far apart from each other and actually connects them, bridges the temporal distance of these two parts interactively. Such synchronization processes take place continuously, but we don't seem to have noticed them often enough or clearly enough. Obviously, the entangled events that we have observed so far are such physical events that confront us with the whole paradox that is not so much due to a falseness of the theory of relativity, but rather to a misunderstood relationship between relativity and quantum theory. Quantum motion does not take place in pre-given space-time-frames, but it constitutes these as if in a spillover for the measuring mind.

I would now like to turn briefly to a mathematical question concerning presence and a step in time that allows us to go back to the basics of human perception. We all perceive what is present in sense awareness. We also speak of "being alert," of "being present," and in English of "presence" and "awareness." This mathematical aspect of the present has to do with computer science. You may have noticed that I have been studying some of the work of the computer scientist Joel Isaacson very carefully for a few years. Isaacson and Kaufmann developed a theory and method in recent decades that they called Recursive Distinctioning. It is a method of recursive differentiation. Now, we have just said that the events of our world all take place in presence. It may therefore not be particularly important to record the various threats of the past linearly and then possibly wind them on the drum of some prayer wheel.

The following occurs with recursive differentiation: a processor operates on the information available to it, which we call input. It detects differences in a certain finite sequential arrangement and calculates a resulting sequence. In short, the processor recognizes input states and uses them to calculate output states. The processor is therefore always in the past. Then it recurs, looking back, diagnosing differences, and developing the resulting conditions. It presents those results to us, to the computer scientists and physicists, and others. The processor therefore takes a step in time by going recursively. It seems that without this one step in time no recursive distinctioning is possible. But this is not the case. We spoke of the light for which time is not binding, and in a few works I even spoke very clearly of movement outside of time. That means that there are material movements which, I say, are stimulated by clear light, but for which time is completely non-binding and non-existent. Let us put it this way: The photon field and, more generally, the boson field contain parts that move entirely without the consumption or flow of time.

This phenomenon, which we have not yet sufficiently understood, apparently has something to do with the religious aspects of science, because it reduces one of Spinoza's qualities of God to a relatively easy to describe process. Namely, if recursion in an intellector processor does not require a step in time, it obviously has to do with self-reference in the present. The phenomenology with which we should now deal is the phenomenology of an intelligent field, which is certainly capable of perception and which, with reference to itself, tends to produce self-referential movement to produce a disaggregation in itself. The clear light, resting in itself, observing itself, falls back into itself, remains in a living state of self-reference, without creating an image of itself. I know it all sounds very crazy, but I cannot now put it any other way. Our search for timelessness, for eternity, for God, is nothing more than an expression of that universal clear light that fluctuates away from time and tries to remain in the state of self-reference, while it moves in time in certain domains, i.e., it stimulates material movement. A recursive differentiation away from time is entirely possible, but it is a very complex process in which separate domains of clear light moving apart from time connect and interact with one another as it appears for the external observer. The clear, self-referencing light, which is self-recurring, i.e., to a certain extent as self-referring and referred to, relinquishes observation, is only reached and understood by us humans, i.e., by our neural system, when we are in that state of mind in which life and death are one. This is the present reality in which the past and the future are folded in and united. I know that sounds very crazy, but it's a living fact. I am talking about these living facts here, and I do not want to argue about what is appreciated and what is not held, but at best I can express good hope.

Chapter 2

What Will Become of Us

Decay of the Huu Men

When you read this, much is already over. The entire text will appear anachronistic. Much of what has been neatly prepared, that for which numerous expert opinions had been obtained, has actually already occurred. It was all prepared within a compliance matrix that is partly understood by a few of us only. Humanity as a whole can no longer exist in its original form. Conflicts are numerous and unbalanced. The extensive communication in an involuted, toppled cognitive system releases a great separation potential. All hidden and repressed conflicts are so intensely pulled into the daylight that only disaggregation of the whole of humanity starts up further development. Of course, some of the most powerful know this. But these most powerful are not easy to identify. They can hardly be located as they handshake each other while they pass the relay to another. No one will admit what is happening to civil societies on every continent. No one, not even the richest, are interested in the fact that public broadcasting stations really know about it. But as all moderators see themselves as important, using the little information they get, they will warmly instruct the sheep in what has to be done with them. By my anachronistic writing, you will see and reflect on what has happened to us. You will remember while reading. At the moment you already have several social fragments that are trying to maintain their stability. Maybe you are somewhere in the middle of warfare somewhere in the Middle East, or maybe you are just lying on a divan in the Middle West and reading this.

A human life is not what we imagine, and it is nothing else either. Because everything we are saying is just another image of thought. Thought, however, the human mind, is just a small thread in the whole moving reality.

I don't know exactly what will become of us. As long as I wrote for the mainstream, I knew better than I do now. But, instead, today I am writing for ufologists, or for people who, like my brother, are greatly interested in ufology. The ufologists, lovers of *Ancient Aliens* are way ahead of us. Because they have dealt with experiences that other people have had that were of an extraordinary nature away from mainstream belief.

When I was studying physics and mathematics, and that's more than 50 years ago, I was pretty sure that physicists had the progress of humanity in hand. We determined how we would go on. We knew pretty well which developments were current, which technical innovations would see the light of day in the coming decades. Today it is clear to me that we, in particular after World War II, have by no means determined these developments ourselves. I described some of the encounters with extraterrestrial civilizations and our encounter with different species that drove technical change in the book on the nuclear journey through time. It became clear to me that those influences of extraterrestrial people and other living creatures were enormous. The effects are incredibly strong, socially relevant and politically dominant, especially now. They will change the systemic structure of our societies from scratch, even overturn them.

Human society, and it is almost beyond me to speak of a community, is violently broken up. No conspiracy theory is required to describe this distribution of social and unsocial fragments. Rather, the ongoing social disaggregation is a process that we have not yet understood sufficiently. But some of the upcoming changes are already being noticed and are discussed in various forums. As is well known, there were philosophers in the 19th century who distinguished community and society from one another. Community refers to groups of people in whom connections and unity seem to be secured, as it were. The philosopher Walter Seitter wrote in an internal protocol from February 2020: "*Their prototype is the family.* Society is *a group of people who sometimes coincidentally live together in a room.*" When it comes to our contacts with extraterrestrials and their spacecraft, we encounter numerous groups that are artificially separated from one another, in particular by the regulations of the secret services. We are then not dealing with communities, but with many small collectives, strongly-bound working collectivities, which ideally can come together to form societies that are then to be located somewhere near the ufology interest groups.

It is very difficult for most of us to understand today when we hear that there are people, whether connected to NASA or not, who have lived and worked on distant planets. It seems completely unbelievable to us that a group of people, along with some aliens, have made their way to a star that is about 40 light-years from Earth. But there are these different, relatively well-separated groups. The people who belong to or formerly belonged to these groups have very little to do with each other. When we talk about a group of NASA members who have made their way to Alpha Centauri or Sirius, these people have very little to do with the members of NASA we know and as get to know on television. They form disaggregated groups whose interests and actions are strictly separated from one another and who can hardly communicate with each other. They are separated from each other by capital and political power. So, when we hear that Ben Rich said that Lockheed Martin Skunk Works can bring the ET home because they already have spaceships that can use extraterrestrial technology and scientific understanding, we should take that seriously. It happened. Such a group of people who have approached Mars, who even in some small frame tried to partially resettle it can certainly coexist with a second group that is following a completely different path of development here on Earth. These are the people at home who are learning now how to set up a habitat on Mars, where to plant lettuce and how to do a little terraforming. They learn how to set up their lives out there badly and slowly.

I am now concerned with identifying the most important groups, namely those of us who will live on our earth, on this currently still-beautiful stormy planet, and also those who leave this earth, who will give up and stay elsewhere. I would like to describe the appearance of these groups with minimal sociological effort and make them plausible.

A lot has happened since Admiral Philip Corso began managing those small component parts left over from the UFO crashes and Intel was commissioned to reproduce some peculiar circuits. It is clear to me, personally, that the flat screen was already available in the late 1950s. You actually have to ask yourself why it took so long for certain devices to reach the market. If you turn on your computer today and you see these rotating spheres in the center of the screen, I may not have to tell you that this is the appearance of certain vacuoles of trapped antimatter that the Earthlings found on various power supplies of the Ebens. Development has progressed so rapidly in the past 50 years that we can literally read out in the present the cross-section of the coming development, i.e., today in the here and now, on our screens. The developments are rushing.

Sitting in your home office, it may happen that you just want to translate a piece of text into English; a message from Google arrives at the bottom right of the screen. It explains that you need to inform yourself about a whole lot of events, about the legal situation, because they may soon no longer be able to use the Google Translate translation machine because some legal situation has just been encountered. In order to understand this alleged legal situation, however, a thorough legal study is probably not enough. You are overwhelmed by a power whose strategies and social structure you do not know and cannot know. Maybe you have the courage to send feedback: "Who actually came up with the idea of taking screenshots on the screens of users' computers? Who advised you of such attacks? Why do I, a user, have to prove that I am not a robot while the operators of the Internet machines hide behind algorithms?" Why are they allowed to do that? Who allowed that? Why has no legal system protected the so-called users from the criminal machinations of the Internet giants? You will surely have observed that the inside of your computer is by no means yours. You do not have all the information that comes in to you from the outside and is, I would say, not only made available to you, but imposed on you. The law was defeated many years ago. Unfortunately, the prosecutors overlooked it.

Eisenhower knew about things that would come. Aliens had warned him of the consequences of alien technology transfer and rapid development, and he too tried to warn civil society leaders that such developments would have dire consequences. I want to go into some of this as I try to classify the coming human development. Someone who sends feedback to Google is not just a little bit crazy, but belongs to a kind of people that I call humanizers. The humanizer is a person who loves people, who wants people from flesh and blood with their minds as we know them to live on and not die out or perish in a nuclear disaster. The humanizer is not only a womanizer, they are also manizers. He is a person who has neutralized the gender gap. She is by no means on the side of those women who emancipate themselves and ensure that whistleblowers like Julian Assange or Edward Snowden go to prison. Humanizers are called conspirators by the mediocre establishment.

But there are still other activities that allow us to classify civil society – and I mean the loose association that happened to come together. In a way, I take an "*actors view*" of how we sociologists do this. I still consider people to be responsible for their actions. It is typical of the young humanizer that they do not hide behind robots, behind machines, behind algorithms, but insist that people are responsible, not only for their own actions, but for human actions in general. That is, if something goes wrong next to me, then I'm also responsible and not just the person next to me. This is how the humanizer acts. Many humanizers I know are artists.

There are other fantasy methods of action besides humanizing. I call one of them creaturing. Here again, there are several possibilities. One possibility that is relatively close to nature is the genetic engineering of humans. With biological genetic engineering we create new types of people. Still another type of manipulation creates a hybrid human that is partly robot, partly human. The last actors in this spectrum, who now result from neither humanizing nor creaturing describe a group of robots and "intellectors," pure AI life-forms. The most advanced of these is an artificial living actor (ALA). This last group is a special kind that comes about by embedding a biochemical intelligent process in the universal life process. It represents an AI that we understand as much and as little as we understand life itself. We can only wonder about the ALA's lives. But that's basically just honest and nothing new. We shall have to confess that we cannot entirely understand creation, even if we appear to be the creators. The existence of living things will always amaze us. This is the reason for founding religions on the Earth and elsewhere. This ensures the existence of religions and of all possible levels of enlightenment in the entire universe. Some may not like this, but it seems to be the case.

But, here now – it is April 2020 – how will we go from there to where?

Too Big to Fail

> The poor guys are driven by greed. They will no longer find a binding language. They invent a pandemic so they can fight it afterwards. Their suffering is immeasurable. It's the lies that hurt huu men so much.
>
> ET

Inequalities in resources and values, as there are sales, land ownership, number of employees etc., in society and the economy are accompanied by mathematical formulas, statistical distributions, such as Pareto, power law, and a number of similar quantities [10, 11]. Associated phenomena are called scale invariant. Consider the distribution of medium-sized businesses by number of employees in Germany 2018: 81% had less than 5, only 2% had 50 or more. It is not only in Germany, but everywhere in the so-called western world, that we find distributions of company sizes in terms of the number of employees roughly the same as the one described above [12].

> EU Commission President Ursula von der Leyen wants "an answer that was born out of the crisis, but that must invest and serve in the future." She stated: "We should now courageously take the chance to create a modern, clean and healthy economy to build up."[3]

But what will such a *modern, clean, and healthy* economy look like? Sociologists, economists, and political scientists have known for almost 200 years that it is easier to close small businesses than large ones. It is possible to consider a lot of small "quantum crumbs," so to speak; we can remove one crumb after another and think carefully about how many crumbs we want to remove. This is not possible with large companies, such as airlines or oil companies. They are too big to simply be removed: *too big to fail.*

We find Heinz von Foerster's *A Discussion on Zipf's Law* [13, 121ff.] recorded at the 1952 Macy Conference, then called the Ninth Conference on Cybernetics, Josiah Macy, Jr. Foundations [14] held on March 20–21, 1952[4] in Hotel Beekman, New York:

> "Now, I asked for this kind of universe the following things – and this is a particular assumption which is made, and I would like to call this particular assumption for lack of a better word, the generalization of the cosmological postulate, and what I mean by a cosmological postulate – which was the famous idea pointed out by Eddington, Millikan, Jennings, and so forth – is that if we look at the world, the world should, from all points of the universe, look alike; that is, wherever you put an observer in the universe he should see the same kind of universe, the same distribution of stars and nebulae. Let us ask

[3] In a radio interview with the Austrian radio station Oe1 on May 14, 2020.

[4] The Macy Conferences failed to reconcile the subjectivity of information (its meaning) and that of the human mind but succeeded in showing how concepts, such as those of the observer, reflexivity, black box systems, and neural networks, would have to be approached in conjunction and eventually overcome in order to form a complete working theory of the mind [WP]. The Macy Conferences were discontinued shortly after the ninth conference [14].

precisely the same thing for this kind of universe; that is, that every observer should see precisely the same kind of distribution. Then we must ask for such a distribution free play, which actually does the job."[5]

In the meantime, Zipf's law, the power law, and related issues of scale invariance play a significant role in work sociology, business administration and economy. In many studies, they provide initial approximations to the empirical quantitative data. The figure below shows the application of such a law to the time series of annual sales of the hundred so-called largest companies in the world [15].

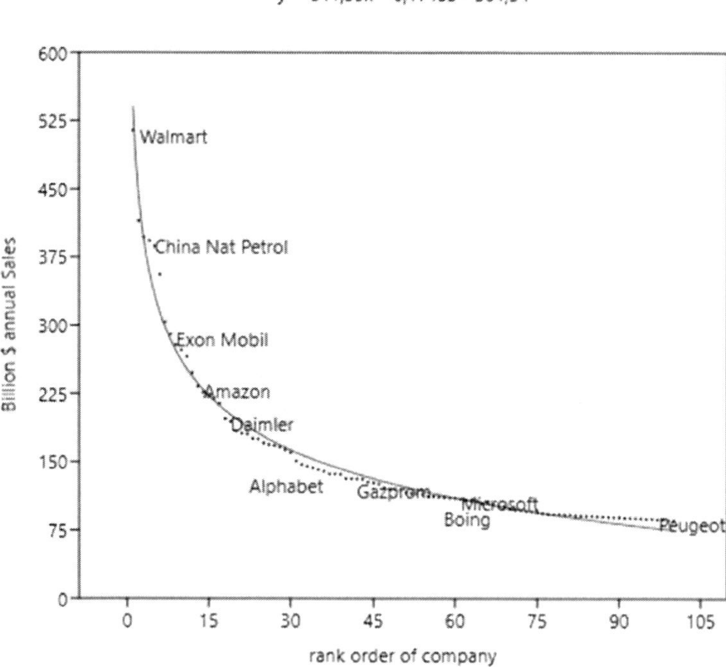

Figure 1. How big are the internet giants?

[5] Heinz von Foerster was the youngest member of the core group of the Macy Conferences on Cybernetics and editor of the five volumes of *Cybernetics* (1949–1953), a series of conference transcripts that represent important foundational conversations in the field. It was von Foerster who suggested that Wiener's term "cybernetics" be applied to this conference series, which was previously called Circular Causal and Feedback Mechanisms in Biological and Social Systems.

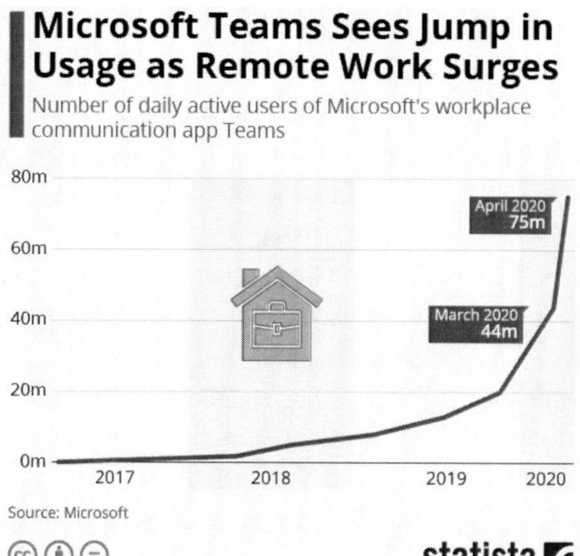

Figure 2. Daily active users of teams.

The numbers on the cited Wikipedia page are said to come from Fortune magazine, which, in June 2019, included the Fortune Global 500, the 100 largest companies in terms of annual sales. There is often talk of Internet giants, but, as you can see among the oil giants, these colossuses are not that huge. The company Alphabet, of which various departments belong to Google, is in position 37. Microsoft takes the 60th place. These ranks clearly existed before the coronavirus crisis. Has the situation of the giants improved in the meantime? [16]

But is that true at all?

L. Rabe said: "In 2019, Alphabet, which is the parent company of Google and other companies, generated sales of approximately $ 161.86 billion. Sales increased by around 18 percent compared to the previous year." [17]

Wikipedia has line 37 allegedly based on [18]:

37 *Alphabet* Internet *$136,819 revenue* United States

Forbes Global 2000 [19] has:

Computer Services *Alphabet* United States *149.7 assets*

J. Clement [20] wrote that "statistic displays Google's operating income from 2013 through 2019. In 2019, the Internet company's operating income amounted to 41.67 billion U.S. dollars. Google is the main revenue generator of online business conglomerate Alphabet."

It seems we have lost a common, binding language. We have no definite terminology for that which allegedly counts most! The whole thing smells of pushing. Nobody should know how high a company's annual turnover actually is. In the penal codes, as I understand, differences in magnitude of such size are well and gladly connected with 10 years' imprisonment. Apparently, statistical lies are mistakes and are not as important as lies in the criminal court. Why and how are people's minds jumping from sales to revenue, from revenue to assets, from there to income, to turnovers and so forth. These numbers are humbug. They do not justify any reliable statistical approach. But that's not all: The swindle is much larger.

OUTDATED INTERIM REPORT: THE CORONAVIRUS PATTERN – A DRAMA

At the time of access (April 9, 2020, 11:18:44) exactly 14,833 coronavirus deceased were counted in the USA; and let us be aware that, when you read this, it is all over and long gone. You might be saying that those were lucky times then! But now, a coronavirus patient is not necessarily diagnosed with COVID-19. With a population of (currently exactly) 330,567,891 this results in around 4.49 deaths per 100,000 inhabitants. Numbers above zero have been accumulating in the United States for five weeks. That's what the statistics say. On December 1, 2017, however, there were already 61,000 influenza deaths in the United

States. At that time, nobody stayed at home. All markets were open. The city administration did not form a "deployment team." The politicians did not restrict freedoms and did not make "flu laws." The car and aircraft manufacturers worked without exception, including the small nail salons. The musicians played neither moody nor exhilarating songs. Everyone did what they felt like. We may have asked: How do people act? And I'd say:

- They don't act knowingly and
- they don't act ignorantly.
- People don't act consciously and
- they don't act unconsciously.

Maybe some say that is impossible. In science, at least some take it for granted, that we assume that we can acquire a whole knowledge, at least not just some half-knowledge. When we forecast the weather, we want statements in the form of "tomorrow it will rain" or "tomorrow it will not rain" or we want to know how many hours it will rain, or what we can expect. Then we make probability statements. We also count such statements as scientific. In this sense, our knowledge of the occurrence or non-occurrence of uncertain events resembles a half-knowledge; yet, our knowledge of its probability of its magnitude is not half-knowledge. The probability that in 600 throws of a die, the occurrence of the 6 on average 100 times is "certainly" one sixth, and we add to that a standard deviation. We know that.

Hence, I simply repeat: We act neither knowingly, nor ignorantly, and we act neither consciously, nor unconsciously. Rather, we act in a pattern that we feel or sense to a certain extent, a pattern that we conform to. In the simplest case, we adapt to an agreement. Two interlocutors can agree on what you will do next. Together, they examine which action they think is sensible in relation to the achievement of a specific goal. In a sense, it is impossible for a single person to act completely independently and alone. Rather, they always act against the background of their experiences, which are social experiences.

I think it is quite expedient to start with the approach of James Samuel Coleman [21, 22], who made it clear that we act collectively, that we pursue our interests, that we are sometimes interested in the interests of our cooperation partners or in the interests of our opponents, that we exercise controls over events and controls over interests, etc. This concept certainly describes human action as collective action within a very large pattern of experiences and exchange relations. Now we can assume that there is a special case in the currently dominating crisis, the so-called coronavirus crisis. We were previously convinced that we had a relatively reliable knowledge of the past, that we had something like history and that we knew about it. In any case, before the Second World War, we were of the opinion that we could say very little about the future, have no precise knowledge, that we were not entirely knowing and not entirely ignorant in the above sense, but at least we were more ignorant than knowing. These findings have changed. We act according to our forecasts.

Large parts of the population are currently barracked. We cannot leave the house. Business people and advertisers, including politicians, are largely restricted in their mobility. The civilian population appears to be experiencing a serious illness, a viral infection like a nasty surprise, so it seems it is best to blindly trust the decisions of the politicians and the measures they take because we don't want to get sick. Everyone wants to survive. And politicians now have a successful motive: We have to protect the very old, our good old people who raised us and worked for us for a lifetime. That sounds very good, but it only appears to give a real motive. However, it defines particular arrows of interest within the collectivities.

Hardly anyone in the civilian population would believe that the coronavirus crisis had been foretold many years ago, that it was planned by a few, and that collective action was predicted. Even the decrees that were supposed to help deal with this crisis had long been prepared and were ready for takeoff. The US secret services and the liaison officers of the deep governments have given the respective state governments sufficient and partly inadequate information. The information quickly disappeared, was "for eyes only" and entire hard drives were shredded. But nobody planned to do anything bad, certainly not.

People do not act consciously or deliberately in all cases. It is enough that they suspect a scenario, that they suspect something that fits in with the zeitgeist. They can then fit into the image of their hunch. They become executors of an imagined notion of the zeitgeist. In fact, there are "coronavirus" scenarios; pandemic scenarios, according to a Rockefeller Foundation report:

Scenarios for the Future of Technology and International Development [23]

"An aggressive virus would spread worldwide, resulting in 20 million deaths" which amazes the moderator [24] with how exactly the reaction patterns of governments around the world were predicted. Is it a coincidence or maybe there is more to it?

Johns Hopkins Bloomberg School of Public Health is currently the most widely-used standard source of global health information; originally named the Johns Hopkins School of Hygiene and Public Health, the school was founded in 1916 by William H. Welch with a grant from the Rockefeller Foundation [25]. The school was renamed the Johns Hopkins Bloomberg School of Public Health on April 20, 2001 in honor of Michael Bloomberg (founder of the eponymous media company) for his financial support and commitment to the school and Johns Hopkins University. Bloomberg has donated a total of $ 2.9 billion to Johns Hopkins University over a period of several decades.

The school is also the founder of Delta Omega (est. 1924), the national honorary society for graduate training in public health. The Bloomberg school is fully accredited by the Council on Education for Public Health (CEPH). Page 18 of the above report states:

> "A world of tighter top-down government control and more authoritarian leadership, with limited innovation and growing citizen pushback. In 2012, the pandemic that the world had been anticipating

for years finally hit. Unlike 2009's H1N1, this new influenza strain – originating from wild geese – was extremely virulent and deadly. Even the most pandemic-prepared nations were quickly overwhelmed when the virus streaked around the world, infecting nearly 20 percent of the global population and killing 8 million in just seven months, the majority of them healthy young adults. The pandemic also had a deadly effect on economies: international mobility of both people and goods screeched to a halt, debilitating industries like tourism and breaking global supply chains. Even locally, normally bustling shops and office buildings sat empty for months, devoid of both employees and customers. The pandemic blanketed the planet — though disproportionate numbers died in Africa, Southeast Asia, and Central America, where the virus spread like wildfire in the absence of official containment protocols. But even in developed countries, containment was a challenge. The United States' initial policy of 'strongly discouraging' citizens from flying proved deadly in its leniency, accelerating the spread of the virus not just within the U.S. but across borders. However, a few countries did fare better – China in particular. The Chinese government's quick imposition and enforcement of mandatory quarantine for all citizens, as well as its instant and near-hermetic sealing off of all borders, saved millions of lives, stopping the spread of the virus far earlier than in other countries and enabling a swifter postpandemic recovery."

This scenario is by no means alone. There have also been preparatory high-level workshops that have been carried out to determine whether scenario-planning techniques provide useful insights into the future impact of ICT on development and the policy choices that governments and development agencies are currently facing. There are several scenarios [26], such as the Event 201 scenario with the nice blue signatures of Melinda and Bill Gates, who also sponsored the ResearchGate.

The Event 201 Scenario

"Event 201 simulates an outbreak of a novel zoonotic coronavirus transmitted from bats to pigs to people that eventually becomes efficiently transmissible from person to person, leading to a severe pandemic. … The scenario ends at the 18-month point, with 65 million deaths. The pandemic is beginning to slow due to the decreasing number of susceptible people. The pandemic will continue at some rate until there is an effective vaccine or until 80–90% of the global population has been exposed. From that point on, it is likely to be an endemic childhood disease."

You don't have to assume anything bad of those people who are often counted as "the elites" without further consideration. They only play with these thoughts. But there is an enormous glare and an unquestioned self-image up there when we use such scenarios to justify politically and legally relevant actions. A major epidemic has been anticipated for a long time, long before the "coronavirus crisis" emerged.

End-of-the-World Scenario: The Global Pandemic

Coronavirus file: Follow the epidemic investigation

By ELSA DOREY Updated on November 29, 2012 at 12:00 pm. Updated on March 24, 2020 at 2:42 pm: "The H5N1 virus recently reminded us that the world's population is living under the threat of potentially dramatic pandemics. The worst? It would be a mutated bird virus."[6]

On April 4, 2017 Jakob Simmank [30] wrote for *Die Zeit*: "The secret WHO boss is Bill Gates. The most important organization in world health, the WHO, has a problem: it is broke and therefore dependent on donations. Is she losing her independence?" Would Jakob Simmank have posted that if it hadn't been true? We'll go into the identifying control of actors over outcomes of WHO actions.

However, regarding the dangerousness of this unexpectedly emerging new virus, we should take care of the associated numbers. I just snapped

---

[6] The dossier (translated by Google Translate as a file) is French: "Dossier coronavirus: suivez les recherches sur l'épidémie." Par ELSA DOREY Le 29 nov 2012 à 00h00 mis à jour 24 mars 2020 à 14h42. "Le virus H5N1 l'a récemment rappelé: la population mondiale vit sous la menace de pandémies potentiellement dramatiques. Le pire? Ce serait un virus aviaire mutant" [27].

a small piece of information that did not concern coronavirus, but rather influenza.

It showed the dramatic course of influenza in the United States at the beginning of the 2017/18 winter season. At that time, there were 61,000 deaths in the United States in early December. Influenza and pneumonia are among the leading causes of death in the United States; graphics can be viewed on the Statista website.[7]

Let's play around with the numbers a bit. If we use the last value of 14.3 influenza deaths per 100,000 inhabitants as published by the CDC among today's population of the United States, this results in approximately 47,700 deaths from influenza. Regarding the last two years in statistics, i.e., 2019 and 2020, the CDC has become a little more modest in its statements and admits a fairly wide range of fluctuation. The CDC estimates that from October 1, 2019 to March 28, 2020, approximately 24,000–63,000 died from influenza. Taking into account the respective population size of the USA, this would correspond to quotients with values of 7.2 to 18.9 per 100,000 inhabitants.

Comparing Influenza with Coronavirus Pandemic

Perhaps I should say at this point that the formats of such statistics can in no way be regarded as uniform. Death rates are sometimes given

---

[7]https://www.statista.com/statistics/184574/deaths-by-influenza-and-pneumonia-in-the-ussince-1950/

as the number of deaths per million people; sometimes no quotient is used at all. There are no uniform requirements. If we compare the statistics for COVID-19, the real-time counter for COVID-19, which is based on data from Johns Hopkins University, shows the two following images, amongst others:

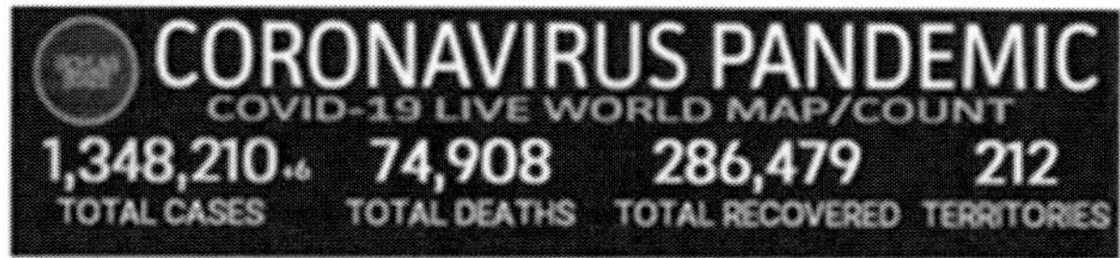

For the USA, I cut out a small icon that refers to April 7, 2020. It looks like this:

You can create such pictures for every day. Let's look at the numbers for specific two days, for example. Then we can create a small trend In doing this, we have to consider the following: Apart from China, the series begins a little over a month ago. Since then, the number of cases, called "total cases" in the real-time counter, has been different from zero. So, the number, especially for the USA and Europe, has only been different from zero for about 35 days; today is 7 April, 2020.

Table 1. Coronavirus cases worldwide and in the USA

		Total Cases	Deaths	Recovered	Territories
April 04	World	1,099,274	59,188	231,957	207
	USA	277,963	7,392	12,283	
April 09	World	1,532,231	89,565	342,111	214
	USA	435,609	14,833	24,125	

The bottom line is that we've accumulated about 11,000 deaths in statistics in the US for about 35 days. Let's compare that to 2017. At that time we had 61,000 people who had died from influenza in the United States until 1 December. These two numbers are therefore comparable in terms of magnitude. I would say that we had more influenza deaths than COVID-19 as counted today. And did those who died die from COVID-19 at all, or are they rather those who died from something else together with being tested positive for COVID-19? Does the WHO test possibly also contain other viruses that are not identical to COVID-19? In any case, influenza appears to be more dangerous if we measure the number of deaths. For coronavirus we get the following disease-specific death rates related to the total population:

Table 2. Death rates worldwide and in the USA

		Death Rate
April 04	World	0.054
	USA	0.027
April 09	World	0.058
	USA	0.034

This corresponds to 4,487 coronavirus dead per 100,000 inhabitants in the USA.

So, these are not numbers that prove the dangerousness of coronavirus. Rather, the spreading speed and the risk of infection appear to be dangerous in COVID-19. COVID-19 viruses are not exposed to the same seasonal fluctuations as influenza. Influenza viruses are known to have a protective shell that melts at high temperatures. Therefore, they can enter the respiratory tract easily and undamaged at low temperatures, i.e., in late autumn and winter. The most important reasons why influenza becomes dangerous in winter can be found at Health Vigil.[8] Corresponding findings are not yet available for COVID-19.

Note update later on 18 May, 2020:

---

[8] Flu Season Months: https://healthvigil.com/flu-time-flu-season-months/.

Today, the numbers show us:

Table 3. Update for May 2020: coronavirus cases and death rate

		Total Cases	Deaths	Recovered	Territories
May 18	World	4,786,963	317,165	1,829,416	218
	USA	1,528,931	90,993	346,389	
		Death Rate			
May 18	World	0.066			
	USA	0.059			

More than two months of counting and manipulating did not change the major observations:

- SARS-CoV-2 is no more deadly than influenza.
- SARS-CoV-2 is currently more contagious.

And still four weeks later the situation has not become worse:

Table 4. June 2020: coronavirus cases and death rate

		Total Cases	Deaths	Recovered	Territories
June 14	World	7,854,192	432,875	3,994,319	218
	USA	2,143,235	117,542	854,106	

In the next spring edition of the *Journal of Space Philosophy*, Yehezkel Dror wrote:

> "1. As of now, the global coronavirus pandemic is child's play compared to the 1918–20 Spanish Flu, which killed more than 50 million people from a much smaller global population – without long-term consequences.
>
> 2. Of the coronavirus, little is known. The numbers of sick and recovered are wild guesses; the duration of post-disease immunity is unknown; no vaccination is in sight; no effective treatment is known.

3. Ergo, no half-reliable model of coronavirus pandemic trajectory can be constructed. Pending more knowledge, only intuitive improvisation is the rule” [29].

In January 2021 we knew more:

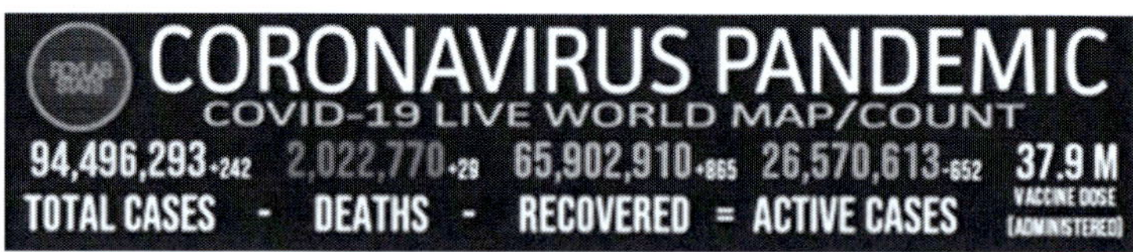

The Microsoft Edge browser displayed a magnificent view on cases and deaths by time series beginning in February 2020. Obviously, in this year we experienced three waves with three impulse functions that triggered the equations of motion.

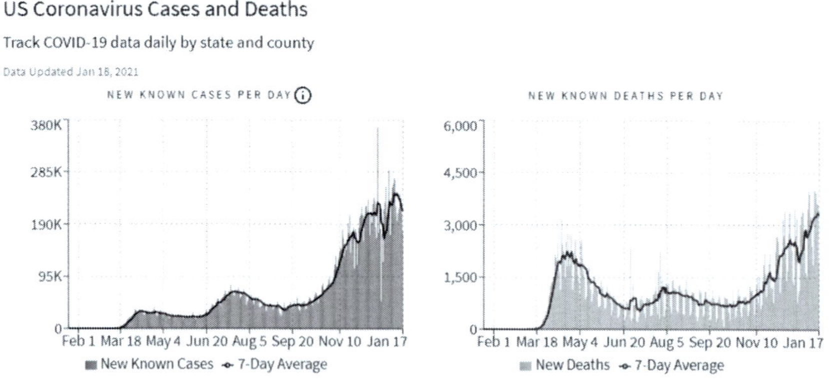

Figure 3. Daily new coronavirus cases and deaths.

Considering Austria, a friend who is somewhat familiar with happenings there said to me, “but Bernd, that is clear anyway. The reasonably alert among us know what goes on.” But why does the government run in such a slipshod way? Noam Chomsky said that the Austrians protected themselves from the worst in a rather selfish manner and refused to help others [30]. The mere existence of such a strategy proves that we are ready to separate from our fellow human beings. Thus, we contribute to the segregation of humanity.

It is about stabilizing new sectors of the economy, positioning the project of information and communication technology (ICT) via circulation and production, and removing a number of old economic structures. It remains to be seen whether the whole thing goes well and goes without war. In any case, even Donald Trump had given in as he realized that the Internet giants were in charge.

Separation Potential

We feel threatened. One of the greatest threats seems to come from outside, from Mother Nature. So-called pandemics are one category of such events, climate change is the next important. The greatest threat is actually our own glare. Too many conflicts, too much lack of trust, too many lies, and too many wrong preferences have blinded us. A colleague on ResearchGate sent me his documented statement.

"To reduce transmission of COVID-19, people in public should stay 2 m away from each other. This is considered a safe distance by public health authorities who promote further measures that include curfews and lockdowns to separate people. In these ways, the incidence of 'social distancing' is keeping pace with the spread of COVID-19. However, as a psychologist points out, social distancing also 'pushes against human beings' fundamental need for connection with one another.'[9] The public health consequences of limiting close human connections may soon include mental health problems such as depression and anxiety, and domestic violence. Gun and ammunition sales have soared in the USA, while in Switzerland justice departments are preparing for increases in

[9] https://news.stanford.edu/2020/03/19/try-distant-socializing-instead/

domestic violence. Community health services will be challenged even further. Multichannel messaging now augments mail and the telephone. People are reaching out to each other in these old and new ways to sustain collective solidarity and redefine the social in their communities" [31, p. 231].

We have done much wrong. The threat now appears to come from outside. It is a fact. Climate change is real. It is not a hypothesis, no theorem, no guesswork, and no mood. Overall, regardless of whether we look inwards or outwards, it is probably the greatest threat to us all. The next in order does not come from the coronavirus crisis, but from a nuclear threat. Main event number three is the nuclear war some are now ready to be waging again. The so-called pandemics, the global epidemics, only come last. We have prepared the situation. The air already contains so many dangerous aerosols and toxins that respiratory diseases are extremely likely. Under normal circumstances our activities are of a social nature. They have always been based on personal contacts, on sensual perceptions, and living relationships. They are carried by the bio-energetic flow of our experience. Currently, this flow, these transactions, these sensory perceptions, are all interrupted by digital data streams and devices. The social dimension has been housed in a network of digital devices. On the television screen in the top left, we can read the main news: "We'll stay at home."

There are still a number of things that we should or should not do under these total circumstances. There is a whole range of behaviors imposed on and demanded from the people of the civilian population, demands which decisively question the personal freedom of everyone. The politicians are trying to determine how far they can go. The political system is testing entirely new ways of manipulation that have rarely been used in history. The current pandemic crisis, if I may call it that, is just the beginning. There will be more pandemics. There will be far more serious illnesses, and the measures taken by the national governments will not be less, but more compelling. Just now someone on the radio says, "We are all afraid of what's coming." Indeed, the next radio show is called "When the digital space replaces the youth center." The potential separation energy in the old science fields we encountered would have known two terms, two types of energy. We know this from physics. First, there is the potential energy that is waiting for conversion, so to speak, and then there is the kinetic energy that is a result of this conversion. These two categories of energy also exist in sociology. A special form of energy is required for each process. We need energy for both attachment and separation. This view is valid in physics as well as in life processes, and especially social events. There is potential separation energy in social life, a separation potential. If a social body has a high *degree of potential separation energy*, it is able to cope with and bring about a large number of separations. Those phenomena that occur in the social action system when separations are being made bring to mind the dynamic aspect of energy, namely the kinetic separation energy. All the migrations, family changes, job changes, regrouping, changes of residence, that appear in this process contain kinetic social energy. If we said that politicians are currently practicing or training something with us, it is primarily the creation of potential separation energy in the social field. Gadgets, PCs, automata.

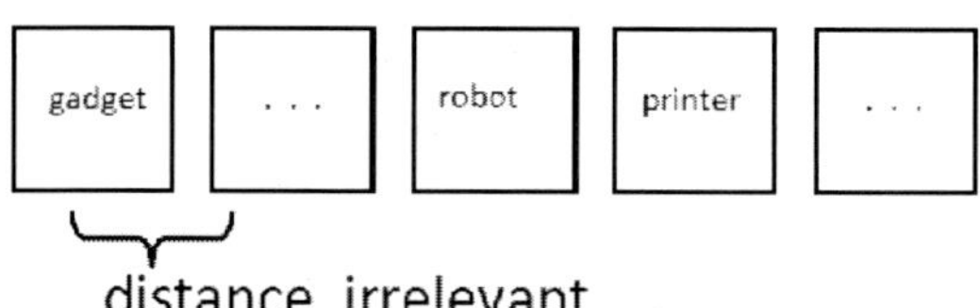

Our society will never again show us those forms that we are used to, the usual separations and bonds we are familiar with. In the past, social formations were determined by ethnic and national circumstances. At the moment, however, these are the sociological exchange systems, the collective action systems, which do not primarily produce geographical and ethnic aggregates, but completely new components, between which very strong, not to be exceeded borders will run. Essentially four or five disjoint social bodies are formed:

1. First of all, these are the people who maintain their original physical constitution. Some of them will probably leave the planet in the foreseeable future.
2. Then there are the people who will survive in hybrid form. They rely on the technical support of robotics. There are hybrid groups that will fight for certain privileges for and against each other.
3. Then there are the genetically modified and renewed people who will carry a new human race into the future.
4. Finally, there are the life forms of artificial intelligence. One of them will be redesigned from scratch, so to speak. It has no resemblance to humans, but has a kind of genetic code and is viable.
5. The last form is a purely technical arrangement of devices that may be reminiscent of human life. In today's highly developed robots, life is only pretended.

In my view, new computer science discoveries will play an important role in the development of the latter two life forms. Microsoft-based robot systems do not have the advanced properties that are needed for AI to become creative. We allow ourselves to be fooled by these devices. These may have extreme computing and intelligence capabilities, but they have no living intelligence and are separated from the cosmic, bioenergetic stream of awareness. Who's acting in our action systems? Currently, the people. But this will change. The computers are already interrupting me and talk to me when I dictate a sentence, and this is by no means a coincidence, it is intentional. The intervals between people and machines are getting smaller and more meaningless. We all like to speak of the geopolitical system. What is this? There are continents. There are national associations, the USA or the Republic of Austria, the many countries with their special systems of action, governments in individual countries, ministers in governments, health ministers. There are non-governmental institutions, such as the Bill and Melinda Gates Foundation, there are the organizations that believe they take care of our health. Traditionally, there are powerful actors, the heads of governments, but also General Director Ghebreyesus, head of the WHO. A an example let us note a few of the players of the times[10]:

Actors	System	Persons	Subsystems
	WHO	Tedros Adhanom Ghebreyesus	Brazzaville Regional Office
	UNO	David Beasley	World Food Program
	Gates Foundation	Bill and Melinda Gates	CHS
	Africa	Robert Mugabe	Zimbabwe
	USA	Donald Trump	Royal Dutch Shell
	GB	Boris Johnson	Royal Dutch Shell
	Germany	Angela Merkel	German Telecom
	China	Xi Jinping	WHO Country Office
	Taiwan	Chen-Yuan Lee	FMPAT
	Global Corporations	Doug McMillon	Walmart
	Science	Tedros Adhanom Ghebreyesus	Bio- and Immunology

So, this is a number of large corporations, super systems, which I have not chosen by chance, but whose representatives play central roles in our drama. The Director General of the WHO is one of the most dazzling personalities. As you can see, I am listing the regional office in Brazzaville as a subsystem and not the headquarters in Geneva, also the name Ghebreyesus in the science category. The last Nobel Prize winner in medicine is not mentioned here, but again the Director General of the WHO. He is actually a biologist and immunologist and is perceived as a very competent person. He is the first African to run the WHO. Why does the UN appear in the table? That's because Ghebreyesus works well with David Beasley, director of the World Food Program (WFP), a UN humanitarian agency. After lockdown all air traffic was paralyzed. Working with the WFP boss made it possible to fully pack WFP relief planes with medical goods, and they didn't start from Brazzaville, but from Ghebreyesus's hometown, Addis Ababa. Robert Mugabe's entry in the Africa category also seems puzzling. Actually, this could rather be, e.g., the name of Michel Yao. He is the WHO emergency coordinator at

---

[10] Meanwhile, some of these actors are no more in their functions or as Mugabe no more alive.

the Brazzaville Regional Office. It's not a coincidence. Of course, I also used a little bit of sociological irony while constructing these tables. Because what are the events that so distract the mighty Director General of WHO? Many people describe him as a friendly, calm, thoughtful person. He looks open and keeps a cool head. Open minds use the potential of diplomacy. How did his political practice come about? Ghebreyesus was Health and Foreign Minister of Ethiopia. That was at the time when China was expanding its influence in Africa. Large projects were funded during this time and the African Union headquarters was built in Addis Ababa. Let's take a quick look back at 2005, when parliamentary elections were held in Ethiopia. There was unrest after the elections. "According to the EU election observers, the election itself was largely peaceful. When it became apparent across the country that the ruling Revolutionary Democratic Front of the Ethiopian Peoples would lose the election, the government stopped the official vote count and declared itself the winner. However, after the opposition accused the government of electoral fraud, Prime Minister Meles Zenawi announced a one-month ban on public meetings for the capital, Addis Ababa. He took command of the military and federal police, imposed curfews and banned any demonstration across the country" [32].

Chapter 3

THE COMPLIANCE MATRIX – A PATTERN OF AMENABILITY

We act in a pattern that we feel and sense to a small extent only. We conform to an invisible pattern. As I said at the beginning, we don't act knowingly and we do not act ignorantly. People don't act consciously and they don't act unconsciously. What does this pattern, which is a matrix of behavior, look like in our case? Curfews are imposed. Protests are prevented. There aren't enough tests. There is no reliable production structure for vaccinations, and so forth. Some pattern slumbers in our brains long before the mysterious coronavirus pandemic breaks out. Each of the above-mentioned corporate actors represents a further complex action system consisting of a large number of actors, events, interests, and controls, which in turn affect the action system. Due to today's electronic communication, that is, the Internet with all its computers, all transactions of the action systems take place to a large extent in the present. The number of communication channels is hardly manageable, the possible conflicts are limitless. The dynamics and possible sets of solutions, which could perhaps once be described using some mathematical systems, say, by eigenvectors of matrices, are no longer clear. If there are any solutions to be fathomed, they are to be regarded to a high degree as technically constructed, highly arbitrary, or technically incomprehensible and mysterious, such as a computer game in chess or Go. If there are any solutions, in a very arbitrary sense, these can only be calculated with the help of the large computer systems of the Internet giants. It would actually not be necessary to search for and find such solutions. But since the techno-systems are there, and the people are extremely playful and adaptable, because we accept any kind of progress, even if it brings on the involution of the cognitive system, and with it the complete annihilation of civilization, we also take it this time, tackle the matter again and try to find solutions within our misery. This will surely not satisfy anyone. Because basically, as long as we humans have a say, it always ends up with some next order Pareto process. The inequality scissors are so horrendous that civilization, with its surprising new means of progress, breaks down our so-called humanity into several strands, which, in turn, incorporate enormous inequalities; and a large part of humanity is destroyed. We are dealing with a really terrifying phenomenon: From now on, it's about killing humanity without being able to identify the culprits. The causes, the intention, the intentional actions are blurred beyond recognition with the help of the technical systems, even completely removed. The cause becomes invisible. This also corresponds to the matrix of mystification. The first major realizations of this crime that no one commits have already taken place. The illegal experiments in biochemical laboratories, the completely crazy decisions within an organization that does not have enough resources to perform its tasks at all – meaning the WHO – have been successfully masked. Some of the poor people to whom the guilt could have been assigned, of whom the misdeeds remain pecking, have fled, or they are so powerful and so well hidden that no one can reach them. The modern dictatorship does not need a dictator.

Most of us currently work in the so-called home office. There is a distance of at least one and a half meters between two of us. The existence of the home office has long been prepared. One may argue whether that was the purpose or not. In any case, with Windows 10 there was always the option to call the workstation a home office. So, we are now working in the home office. There is now not only one meter between two people, but there are the many devices between us. You know the whole game of "social distancing." I don't need to explain that to you. A list of human actions, which are now determined by social distancing, is extremely long, almost infinitely long. With a single decision by the WHO, a state was reached in which basically everyone has to learn, and actually learn, to distance themselves from other people. This breaks the sensual, bioenergetic, contact by body language. This vacuum, this set of new distances, can be used purposefully. I am talking here about a *distance potential* or a *separation potential.* A basic difference system has been installed in society. It induces a transition from the basic space of real-life processes to the formal space spanned by computer processors. It brings on a geopolitical phase-transition *from globalization to digitalization*, and a sociological second order Pareto redistribution process leading to unimaginable further disparity.

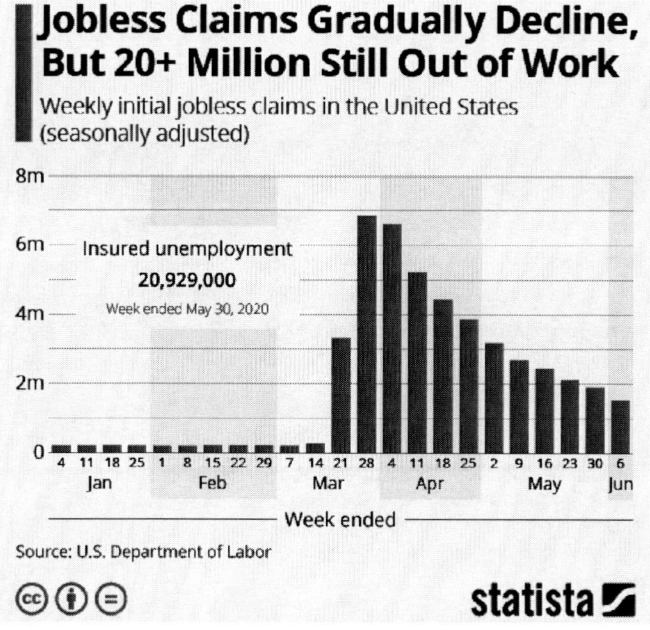

Figure 4. Jobless claims.

"Pandemic Ends Longest Growth Cycle in U.S. History | Weekly Jobless Claims Trend Downward," wrote Felix Richter on June 12, 2020.[11]

"The number of people receiving unemployment benefits also saw a slight decline in the week ended May 30, but with more than 20 million people out of work, the jobs crisis is far from over. When asked about his expectation of how many of the latest job losses will turn out permanent in this week's FOMC press conference, Federal Reserve Chairman Jerome Powell said that it 'could be well into the millions of people who don't get to go back to their old job,' and that 'it could be some years before we get back to those people finding jobs.'

"What had only been a question of time since the coronavirus pandemic hit the U.S. economy with full force became official this week: the United States is in a recession.

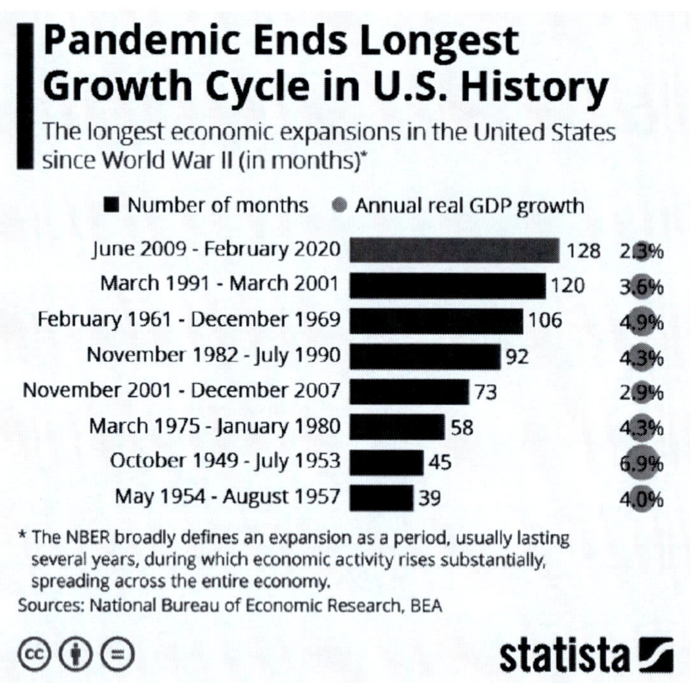

Figure 5. US longest growth cycle.

[11] From Statista Infographics Bulletin - Late Edition on 2020-06-12 21:26.

"On June 8, 2020, the National Bureau of Economic Research (NBER), the chronicler of economic cycles in the United States, announced that February 2020 marked a peak in economic activity, signaling the end of the longest expansion in U.S. history and the beginning of a recession."

"As the above chart shows, the latest expansion was the longest ever, trailed by the growth cycle that lasted from March 1991 to March 2001 and culminated in the bursting of the dot-com bubble. As opposed to many shorter growth periods of the past, the most recent one was characterized by moderate growth. With an average annual GDP growth of 2.3 percent, the longest upturn was in fact the second slowest since the end of World War II."

As the entire global economy falls, the software industry continues to grow, and network managers are making increasingly astronomical gains. The process becomes critical when the great advantages of distributors such as Walmart are to be overhauled. This may be more important than the overhaul of major oil companies. When oil companies and distributors are overtaken in annual sales by IT operations, without a doubt network information will turn out to be more important than the basic human needs of food, drinking, roof, sleep. There is no telling how people will cope with the coming ordeal. When Microsoft and Google have left their ridiculous places at numbers 60 and 30, their legal departments will take over the law of former nations. The legal systems have let civil society down for decades, they are taken over by the Internet and information industries. Science is given new tasks. Consideration and deference to small or medium sized enterprises cannot be expected.

Global networking, the Internet, and software giants will check whether the national governments have done their homework; if they are underrated, they will easily be overthrown. In the foreseeable future, what we have called the rule of law and government will no longer exist. The rich software companies will of course now have enough money available to invest in the investigation and construction of completely new operating systems. Some free operating systems, such as the ones invented by Joel Isaacsson [33, 34, 35], will perhaps be added to quantum computers.[12] These systems are moved not only by binary processing, but by the relation between different forms of logic. Isaacson's system is creative in a deep sense that we first would have to discover. Program-free processors could provide the basis for the creation of life-forms 4 and 5 and they may be useful for the making of life-form 3, the genetically modified and renewed people. They will carry a new human race into the future, probably some type of Ebens.

But, at present we are bound to the compliance matrix. There is our (pre)historical willingness to comply to hegemonic structures and social inequality. Inequalities of resources in society and economy are accompanied by statistical distributions and well-formed mathematical formulas. But experiencing the specific problems involved is more than formulas. May 2020 is now coming to an end and the world is in a crisis that is still largely perceived as economic, but the real consequences are those, they say, are not yet in sight. In the meantime, one hears occasionally of the so-called Pareto principle, at least in the critical radio stations. Sociologists know the importance of this principle, according to which, roughly speaking, many have little and very few have very much. In Wikipedia [11] you will find the name of the economist Vilfredo Pareto, who formulated a law on the distribution of income among the population at the University of Lausanne in 1896. The Pareto principle is also known as the 80/20 rule, the law of the vital few, or the principle of factor sparsity. The Pareto principle states that, for many events, roughly 80 percent of the effects come from 20 percent of the causes. In abstract terms, the Pareto function describes a statistical distribution in such a way that small numbers of high values contribute more to the total value

[12] I haven't said much about these operating systems. Their performance cannot yet be adequately assessed. In my investigations I have shown that Joel Isaacon's intellector systems can not only simulate a quantum processor, but the system has the wonderful, and to a certain extent, lively property with Hegelian dialectics of advancing from one constellation to the next by combining theses and antitheses, synthesizing while overcoming the apparent contradictions.

than the high number of small values. Of course, what exactly that means should be questioned.

Anyway, in May 2020 we heard that Ryanair was closing down and giving up its base at the Vienna International Airport in Schwechat at the end of May. That caused a lot of excitement. Aviation traffic expert Kurt Hoffmann spoke about the dire situation around the world, according to which 25 million jobs worldwide in aviation would currently be on the brink of extinction. In the past week alone, 40,000 jobs had been eliminated in aviation in Europe. The presenter asked what the future would look like for the airline. Answer: Although Lufthansa stumbled, as did AUA, it can be assumed that these airlines will survive, despite 85 percent of all airlines needing jump-starting. Tourism is stagnating with air traffic. Statista showed the following impressive figure:

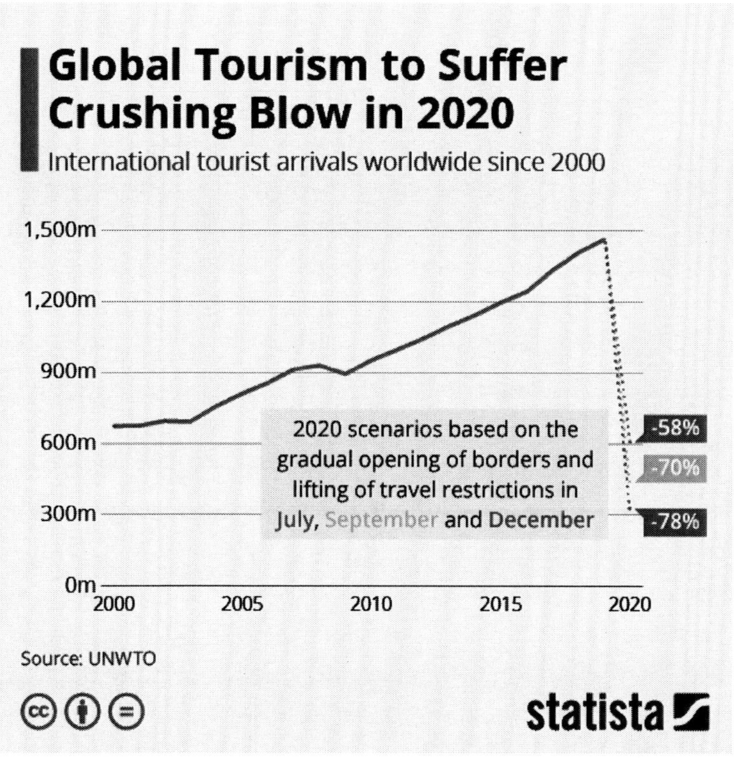

Figure 6. Collapse of tourism.

It is particularly surprising for me to be confronted with these facts while writing a postscript to nuclear space travel. I believe that our view is far too narrow to even roughly estimate the structural and phenomenological changes in aviation. What we see first is that at the moment the air traffic on earth is idle and many airlines, if they start at all, only start with about 25 percent of their capacity. Some large hubs, such as Vienna or Frankfurt, are slowly starting up. But even that is only possible with massive government aid. Ultimately, the money for this comes from the civilian population, from the world's populations. Incidentally, the unemployed pilots cannot switch to other airlines as before, e.g., Emirates or other Arab airlines. These, too, are in the midst of an economic crisis.

Pareto "unequal-distribution" already existed when people hadn't thought about it. The fact that today we have developed a mathematical and statistical theory that is lying ready in front of us, be it the theory of scale invariance, the Zeta, Zipf, or Boltzmann distribution, or something else difficult, that fact started to develop in reaction to our sensory perceptions and observations. We had observed that people compete with each other, that they fight each other, and that their struggles, their disputes, their contests hardly ever come to a sensible end. In the end, we don't even worry if we see our impotence in the face of ongoing wars. It seems completely natural to every young person that weapon systems are constantly being tested, that their effectiveness is continually "improved," which is utter nonsense because weapons are bad for people per se. People kill themselves year after year, and always the state, the church and the legal systems declare the illegal as legitimate. "Staatsnotstand" is not a good thing, it actually mirrors a grotesque, over-inflated weakness of the human being.

Even when we had no theory or knowledge about how the Pareto distribution came about, there were already those capable few who could extract the best of all trees from the forest, and there were the few grand violinists who played the few best Stradivarius violins, there were excellent pilots who flew from one continent to another, and there were many who did not even have a bicycle. There are millions of rivets, billions of good-for-nothing people, and in the crowd, there are a couple of very powerful, superrich, sociologically backward, power-hungry beings. There may be exceptions, but they do not change the reality of the unequal distribution we observe.

These inequalities, which some people noticed centuries ago, did not require higher mathematics to become real. They came from a dark region of ignorance. We can say that in the midst of all the struggles, the results of these struggles are born of ignorance and darkness. It was only our thinking that brought the defects of this dark birth to our eyes, and we are currently faced with them unmediated and directly. It is about getting out of this darkness and moving into the light, which is not only conscious, but in which dark is not separated from light, where conscious and unconscious are one. Only a living being familiar with these two opposite poles is able to live peacefully and at the same time with excellent effect.

It is inconceivable that the current development, which is more of a phase-transition than a development, is not subject to any caution or loving braking action. Why can't any foresight, no foresighted judge, no prosecutor, counteract the catastrophe that awaits us? What is going on there? Well, our entire system is based on fear. Fear is not just a natural component. It permeates our entire social system. The human character is characterized by fear. This fear, which is very often associated with "angst," is fear of not having enough money, fear of not getting a new car, fear of starvation; and all these specific fears prevent our clear view and force a certain way of life in society. We live in tight spaces. But our capital is not money. We don't recognize that either because we are blinded. Our real capital is dependency. Most of us would even be proud to be surrounded by as many friends, acquaintances, and relatives as possible. They think that's a good thing. However, dependence always means that we convince each other of our beliefs and identify with them. Here, again, there are Pareto-distributed beliefs. A few believe in good ideas, many in bad ones. Some consider these and those events important, even if they are completely unimportant, and depending on them, they identify with these ideas. This is our biggest bad luck, so to speak. It may be that some of the operators of the current operations around the coronavirus have even known about the relevance of dependency, and that they therefore consider it beneficial if people have to live at a greater distance from one another – as social distancing, because this also makes identification a little more difficult. There is also a certain deception in this. Identification does not become more difficult, it is now brought about by our devices. To connect with a person in a sensually perceptible proximity, to speak to each other, to touch each other occasionally, to hear one another, to smell, to taste, or not to be able to taste, all of this is much more concrete and affects our genetic code differently and more directly than the dialogic communication with a cell phone or with a computer.

Only those people who have always managed to stand out from the crumpled mass as individuals and to think and work on their own, and do their daily work properly, only these people are now able to do this during social distancing. They keep a clear head. The knotted ones, however, are confused. This now also produces a new type of stratification. Some loners will be released, some groups are made viable in a peculiar, technical, not yet fully understandable way, while the traditionally viable groups are suddenly weak and impoverished. Not a single judge, not a single public prosecutor, was willing to see and fight all of this. Why not? Well, judges and prosecutors are also people who live in addiction and belong to the dumpling of the hardworking and sluggish crowd. This has also meant that some new software gurus can do what they want. I am not even talking about the numerous illusions that they fell for. These people have a completely wrong idea of the social processes that they trigger with their innovations; and I am not referring to Bill Gates. I am referring to many misconceptions about people's lives that are yet to be discussed.

The construction of the Internet and the external processes of thought naturally result in an enslavement of the human being and the establishment of a higher-level networked intelligence that actually brings with it a new kind of life organization, even a new kind of living being. We become – we are made happy – positively happy by placing ourselves in a hybrid, a technical quasi-organism that actually draws its life from us humans.

The involution of the thought process is truly an inversion, as with gloves, if we turn a left leather glove over, it becomes a right one, and it is not said that it is particularly convenient to wear this inverted glove. But there are also involutions and processes that lead to completely new situations that we are unable to cope with because we lack the tools to do so.

There was once a philosopher, Jiddu Krishnamurti, who repeatedly pointed out during his lifetime that technical developments had brought us into a new, threatening situation [36], [37]. He repeatedly asked: If the computers take over human thought processes and actions, what will become of the human brain? Today we can already see that basic skills, such as mastering the basic types of calculations, reading and writing, are difficult for today's children. On the other hand, we have made them dependent on the reward systems of computers, on the five-star economy, and this is a tremendous intervention in the hormonal system of people. This is one of the many serious crimes committed by the software industry and silently accepted by the legal system.

If you want to counter the global crime that is going on, you have to say what you experience. Then you must not bow to the media, the so-called public law, whose employees are all on the drip, and who are promised salaries as long as they meet the growth frenzy of Internet giants, which are not giants at all, and willingly follow the news of the positive spread coronavirus crisis. It is completely inconceivable how lying, how selfish, how greedy some of those people who set the tone are, how they pretend and with what lies they hide their evil intentions. They no longer recognize their own malice, because, in the course of their greed and wealth, they have forgotten and suppressed the information about what they have done.

I find it difficult to talk about these things without insulting specific people by pointing out their carelessness. Rather, I am suggesting that all these misconducts in which we will now drive the addicts, do not happen out of deliberate malice, but rather that everything fits into an overall pattern in which losses are losses, and these are not small losses, but total losses. We will speak of a total war, but nobody knows of it and nobody wants to know anything of it. It is the war of a dictatorship without a dictator. If you follow me, you will surely make yourself an outsider within the large, blinded civil society. If you have the strength to endure this, then follow me! The wage you will receive is clarity. You will not use the word "clear" in every second word in a sentence like the politicians. But the phenomenon of light perception, the view in clear light, will be completely familiar to you, and a matter of course. In this state of happiness, you are united with life itself and if you see what others want to impose on you, but it will not affect you. It won't touch your dignity.

It is the happiness of being united with life, a life that we have always called God, in which we actually experience the solution and detachment from all human and social problems. There is no need to join a religion, so you don't have to become a Buddhist. You don't need yoga or a guru. And, above all, you don't need time for this, because life moves away from time. Time is just one of many products in life. Surely it is not the case that the whole universe is ordered by linear or non-linear time. Time is something extremely vague, which is only constituted in the execution of human thought so far, what physicists think they are experiencing and about which they speak. Life moves apart from time. By stepping out of time you come into life. This does not require a reaction to the coronavirus crisis, but many who know about the phenomenon of enlightenment, who are familiar with the encounter with life, they have always had it with them, and so they cannot really be surprised, even by the greatest of any suffering; and suffering does nothing to solve a crisis.

Life does not learn from experience. Life makes experience. It creates memory systems creatively. But life itself is always above experience. Stand out from the dumplings! Make yourself independent, no, do nothing at all, just realize that they never belonged to them, that they never listened to the banal evaluations! Then the stepping out is done by itself every day. Then there is no me that wants to decorate itself.

Then you live. Nothing else is needed for this, no priest, no pope, no state, no president. Only when you are free of these labels will you see that you have always been free and that you have never needed a coronavirus crisis or any other virus to get any dubious and supposedly healing insights. To be free from all these errors made by humans is light in a bliss that exists far from all medicines and advice and masks. A truly living person has no mask.

Our Internet giants will have noticed that, despite all the networking, our society is divided into many actors and groups. They believe that networking helps. That is not the case. Even the few billionaires are badly integrated. They cannot agree upon relevant issues. The threatening mixture of climate catastrophes and wars that are spreading all over the world will probably make us feel very uncomfortable. That means that we will look for new ways, and as always before, we will go on the run to find new shelter. The sincere poor may survive here and there as people as we have known them so far. But many will be lost. This is perhaps the only fact about which the rich and powerful can achieve agreement.

Chapter 4

FOUR THREADS AT FOURTH TURNING

The online lexica contain terms such as:

- cosmic inflation,
- population explosion,
- information explosion,
- nuclear explosion.

These are all relatively dramatic terms. We may or may not be critical of these denotations. In any case, they reflect the drama of the interrelated contents. Bombs explode, populations exploded until recently, and information is exploding too. There can be no doubt about that. It seems probably less dangerous if we start with terms such as information density and population density, instead of "explosion." We cannot, however, conceal the dramatic effects of nuclear explosions, or the exponential increase in the density of greenhouse gases in the atmosphere, with well-chosen words alone. We are, indeed, having all these explosions. There is this high population density and the great closeness between animals and humans in which pandemics can spread unhindered. There is also this more-than-exponential growth in data flow. Like population growth at the time, this is based on a non-linear main unit or parent population, since information exchange affects not only sets of information sources, but Cartesian products of quantities of senders, transceivers, and receivers. All these explosions do have actors. The actors of cosmic inflation and dark matter are physicists. We might believe that these actors are not relevant. But they are. Cosmology will have consequences for aviation. Then we have the actors responsible for population growth, the people and their politicians. It seems that the electronic boys responsible for information explosion are well known. But that, too, is an illusion. The actors causing the information explosion are distributed in the net. They cannot easily be localized and identified. Then you have those guys responsible for nuclear explosions, most of them military men and politicians. And last, but not least, there are the well-preserved humans, but also genetically engineered humans, artificial humanoids, hybrid humanoids, ALAs and the purely artificial robot intelligences. Considering these as specific groups of actors, it is obvious that these groups are all connected in a kind of predator-prey relation. They hunt or they avoid each other.

The software engineers are interested in helping the needy and disabled people. They provide the necessary interfaces that connect artificial with natural intelligence. This draws sick people into the hybridization cycle. The military, in particular, is interested in helping mutilated soldiers. The neurologists are interested in artificial intelligence. They would like to create an artificial brain. Such attitudes correspond to our curiosity and our research spirit. ALAs will perhaps be protected by all four – humans, hybrids, humanoids and robot systems – since the ALAs represent the most important biological creation of humans. Laws will protect ALAs and robots most, and they will care in some well-known ambiguous manner for hybrid men. The number of sanely preserved humans with high moral standards will decrease, because they will be exposed to all the threats of the planet, the pandemics, the climate catastrophe, and the nuclear accidents. These intact humans won't be very popular. They are rather similar to those who warned Dwight Eisenhower. Yet, the nuclear threat prevailed.

I wish we could see this reality clearly. On an Arkansas evening, September 18, 1980, "a worker slips a wrench out of the hand while doing routine work on a ready-to-fight 'Titan II' rocket. It falls into the silo and tears a hole. This mishap leads to a chain reaction, which threatens the worst-case scenario, the largest nuclear accident to occur in peacetime. The largest nuclear warhead ever built is about to explode. The United States is on the brink of a nuclear disaster. The governor of the US state of Arkansas, during his first term, was Bill Clinton, who later became the 42nd president of the United States. Like everyone involved, Bill Clinton is not prepared for a crisis of this magnitude" [38]. At the end of the documentary was the summary: "Since the beginning of the nuclear age, the United States has built 70,000 nuclear weapons. None of them have ever been accidentally detonated. We owe this to the skills of our weapon designers, whose recommendations were ultimately implemented, and the courage of our military personnel. But we were also lucky, just lucky, and the problem is that luck will run out at some point. Nuclear weapons are machines and ultimately something goes wrong with every machine that was ever invented. ... No matter how many checklists you have, someone will present you with an unexpected problem. ... Nuclear weapons will always carry the possibility of unintentional firing. It will happen, maybe tomorrow, maybe in a million years. But it will happen."

That statement, "the problem is that luck will run out at some point," is statistically untenable, but is quite understandable sociologically. Perhaps we force our good luck to run out. So many important people claim that we are at war. Ghebreyesus said we were at war, Steve Bannon said the same thing. Although Donald Trump removed White House Chief Strategist Bannon from office in 2017, Bannon's rhetoric was reflected afterwards in the statements of the White House. The Fourth Turning has become an obvious legitimation to stir the war drum: "We are at war. America is at war." But what is the reason for such a widespread feeling that we – together with this COVID-19 crisis – are at war? We are not honest with ourselves. We are repressing our perception of reality. This is the perception of the enormous density of information that we can now actually call an explosion. With the number of users, devices, and gadgets, the number of communication channels and possible transactions increases to an unforeseen and sociologically devastating extent. We can assume that there is a certain probability measure for potential conflicts associated with each communication channel. However, as long as these communication channels only represent some potential and are not used, perhaps simply because they do not yet exist, this informative measure of probability is ineffective for conflicts. A realization of the conflicts is only possible through the implementation of the communication network. Potentials become real conflict situations. This entire process provides the actual basis for the feelings that we are not fully conscious of and which suggest that we are in a kind of war. The competent decision makers in information technology are well aware of these processes. As we have shown before, they have even created scenarios in which this situation is examined. Now, to a certain extent, all conflicts hidden by the information technology and software industries have been completely hidden behind the coronavirus crisis. The situation is particularly dangerous because, according to a certain theory, the United States is said to be actually entering a war episode. Unfortunately, this theory seems particularly important to both the Trump administration and key people in the ecological movement. The so-called Strauss–Howe generational theory states that after every saeculum, acrisis recurs in American history, which is followed by a recovery.

> “The Strauss–Howe generational theory, also known as the Fourth Turning theory or simply the Fourth Turning, describes a theorized recurring generation cycle in American history and_global history. … Each generational persona unleashes a new era (called a turning) lasting around 20–22 years, in which a new social, political, and economic climate exists. They are part of a larger cyclical ‘saeculum’ (a long human life, which usually spans between 80 and 90 years, although some saecula have lasted longer). The theory states that after every saeculum, a crisis recurs in American history, which is followed by a

recovery (high). During this recovery, institutions and communitarian values are strong." [39, p. 125]

In the current crisis, all origins of violence are hidden. There is no immediately identifiable cause, nor is there a dictator. There seem to be several dictators, but their dictation is also difficult to make out. Is it possible that we are dealing with processes of resilience? Ingolfur Blühdorn, Director of the Institute for Social Change and Sustainability at the Vienna University of Economics and Business said in a radio interview on resilient societies:[13] "The term resilience is interesting because it shimmers. The term resilience actually means that a system is able to stabilize again after a shock or after severe irritation. That is interesting. Since if that is the case resilience does not mean change. The concept of resilience could be compared with the concept of transformation with good reason. Either modern societies are transforming or they are resilient. However, if we take this ambiguity of resilience, we can very well say that corona strengthens the resilience of the preexisting system, namely the pandemic reaffirms the preexisting system with all of its economic, political and social unacceptabilities."

In fact, however, we are in the midst of a phase_transition in which all social subsystems are under the enormous pressure of profiling and differentiating. All subsystems, whether economic or political, are exposed to a high-energy Pareto second-order process. Therefore, when the system returns to its traditional structural and qualitative form, we must expect unprecedented inequalities and class-like struggles. The very fact that we cannot identify a direct cause of the processes means that, if the nuclear war systems can be avoided at all, there is no other option in the system than to enforce completely new social and political boundaries. The system disaggregates at a higher, previously unknown level. The essential aggregates are already there. These are:

---

[13] Blühdorn gave an interview to Austrian Radiokolleg (broadcast college) Oe1 on resilient societies in July 2020.

- preserved humans,
- hybrids,
- genetically modified humanoids,
- artificial living actors, and
- artificial intelligence with the intellector property.

The now significant reduction in social interactions and the politics of social distancing lead to a clear distinction, shape, and delimitation of these social aggregates.

The actual design processes, the sociological drivers of this new aggregation, are extremely directional and are due to the detailed processes of information technology. Human society reacts to the pressure of the information explosion with differentiation. We have already discussed some of the explosive expansion processes that have now been triggered in the IT industry. For completeness, we have to consider the following. In accord with Wikipedia contributors, we can say that "the information explosion is the rapid increase in the amount of published information or data and the effects of this abundance. The earliest use of the phrase seems to have been in an IBM advertising supplement to the New York Times published on April 30, 1961, and by Frank Fremont-Smith, Director of the American Institute of Biological Sciences Interdisciplinary Conference Program, in an April 1961 article in the AIBS Bulletin [40, p. 18]. The world's effective capacity to exchange information through two-way telecommunication networks was 0.281 exabytes of (optimally compressed) information in 1986, 0.471 in 1993, 2.2 in 2000, and 65 (optimally compressed) exabytes in 2007" [41].

Over time, some information technology artists seem to have realized that the idiotic all-present visualization of information not only destroys civilization, but also its history. In the end, the Wayback Machine was added to the Google archive out of necessity rather than virtue. "The Wayback Machine is a digital archive of the World Wide Web, founded by the Internet Archive, a nonprofit organization based in San Francisco. It allows the user to go 'back in time' and see what websites looked like in the past. Its founders, Brewster Kahle and Bruce Gilliat, developed the

Wayback Machine with the intention of providing 'universal access to all knowledge' by preserving archived copies of defunct webpages.

Between October 2013 and March 2015, the website's global Alexa rank changed from 163 to 208. In March 2019 the rank was at 244." [42].

Table 5. Wayback machine growth

Wayback Machine Growth	
Wayback Machine by Year	Pages Archived (billion)
2005	40
2008	85
2012	150
2013	373
2014	400
2015	452

In this new context of social structures, we can ask who will be able to acquire and take advantage of time travel technology and who will start the journey to the Pleiades and other planetary systems that house human societies. And, unfortunately, I have to give you a disillusioning answer here. The intelligence within the so-called cloud and modern information technology is by no means to be assessed as large, but rather as insufficient for a comprehensive understanding of the nuclear journey through time. The well-preserved humans are also unlikely to become those people that are particularly favored by the system. Furthermore, both computer science and modern, overgrown mathematical physics are almost too stupid to understand phenomenology along with the algebra of nuclear time travel. The compromise of the TR3B Astra aircraft clearly shows us that we do not understand the relationship between strong interaction and the dynamics of space-time sufficiently well. Therefore, we have used some of the old, rather obsolete technology of our ancient aliens as it came up to us from the ancient Indian scriptures. This technique is out of date. The new physics technique requires a thorough revision of theoretical physics, especially that of quantum phenomenology of motion that has not been understood at all. My books were the first successful attempts to make the theoretical physics of nuclear time travel understandable. I hope that there are a few who can understand and carry out the matter.

The times when the interactions between the Ebens and the engineers of the United States were close enough to accelerate the technology transfer in terms of time travel are long gone. The extraterrestrials understood that communication with humans should be handled sparingly. Nevertheless, human greed and national diversity within humanity have led to such rapid progress that the consequences have become catastrophic for us. We did not understand the technology of space travel and its journey through time. We only appropriated the information technology and the associated material technology. But that is not enough. Without space travel, humanity now is defenseless against the pressure of information_technological changes. The only responsible people I have noticed who encourage easy answers and full outdoor adventures belong to the Kepler Institute of the United States. It is a handful of people who I think can make a significant contribution, especially Bob Krone. However, he is a very old man. It would be desirable for talented people to follow him, and to make his plans and the ideas of his friend Salina a reality.

Richard Boylan, a behavioral and social scientist, psychology professor emeritus, and anthropologist of star cultures, more than ten years ago made an inquiry about "classified Advanced Antigravity Aerospace Craft utilizing back-engineered extraterrestrial technology." Boylan works with the "cosmic generation" of advanced, psychic, and space-aware Star Kids and Star Seed adults. As a Star Nations' Councillor (Ambassador) for Earth, his assignment is to advance knowledge, understanding, and a good relationship between Earth and the peoples of space.[14]

Boylan is neither a physicist nor an aviation engineer, but, in any case, he feels responsible for the unusual experiences of people who have come into contact with the phenomena of exoanthropology. His studies list the aircraft of which at least some use the antigravity technology: "the

---

[14] Quoted after http://www.drboylan.com/.

Northrop Grumman B-2 Spirit Stealth Bomber, the F-22 Raptor advanced stealth fighter, and its successor, the F-35 Lightning II advanced stealth fighter; the Aurora, Lockheed-Martin's X-33A, the Lockheed X-22A two-man antigravity disc fighter, Boeing and Airbus Industries' Nautilus, the TR3-A Pumpkinseed, the TR3-B Triangle, Northrop's 'Great Pumpkin' disc, Teledyne Ryan Aeronautical's XH-75D Shark antigravity helicopter, and the Northrop Quantum Teleportation Disc."[15] I have studied his explanations in detail and slowly brought them to my mind. His report not only shows the physical situation rather precisely, but also makes clear the faults and the enormously deep rupture line between humans and those extraterrestrials whose technological knowledge we believed we could easily take on. It is this break line – our greed – from which today's conflicts have arisen. Of course, these are not just physical and sociological facts, we are also dealing with an enormous variety of ethnological groups on Earth that can be traced back to the presence of extraterrestrials. Our life today is based on a sparse but long-term presence of extraterrestrial beings.

GOING FORWARD BY LIVING BACKWARDS

Time travel technology – I would like to call it *looking glass technology* for a moment – is underdeveloped compared to biology and information technology. The development of aviation could not keep up with the software industry at all. American secrecy practices are to blame for this. The conflicts in the area 51 between the military and the Greys naturally did not promote communication between earthly and extraterrestrial scientists. But yet we can assume that space travel and in particular time travel technology have developed considerably since engineering alien craft back in the 1960s and 70s. Only civil society knows nothing about it. Civilians also know little or nothing about nuclear accidents. Governments and their research institutions are hiding

---

[15] https://www.metatech.org/wp/ufos/secret-government-anti-gravity-fleet/.

their very negative experiences, but they are also hiding their very positive experiences, namely those that brought unusual technological advances, especially where they are rooted in extraterrestrial technology. Unfortunately, the many valuable insights from special research departments of black intelligence projects were not linked with another and they still are not. Thank God there are some brave people who contrast the secret projects with public projects having similar names today or compiling many such historic names. The Pegasus project is one of them. Meanwhile, our findings lie in front of us like a huge heap of crumbs. Of course, there are small elite units in naval research and aviation research where the many crumbs are collected. But despite these signs of deterioration in aviation and space travel, the concept of the *looking glass* and the so-called Alice-code have remained intact.

Dan Burisch and Bob Lazar were young men when they were used by the secret services in special research and development black projects. They brought to light some of the most important findings. Despite the thrift and modesty with which they originally brought up this content, the mainstream didn't believe them; even today, many shake their heads when confronted with their original interviews and documents. These actually concern time travel technology.

There are now many bloggers and websites that record the experiences of the two. A first glimpse can be gained in the section "*Forbidden Technology Part Two, Project Looking Glass*" [43] [44]. This is a conversation between Kerry Lynn Cassidy and Dan Burisch. In it, Burisch then reflects on Bill and Will Uhouse's earliest experiences with the looking glass technology. If someone tells you that it is all nonsense, then I advise you to remain calm. It is not nonsense. The thing is real. We'll take a closer look at that.

In *Project Aquarius and the Looking Glass Project at Facility S4 at Area 51* [45], Burisch describes the second floor or "4-2" that was known as "Alice's Floor." Six years after *Alice's Adventures in Wonderland*, Lewis Carol had written the novel *Through the Looking Glass, and What Alice Found There* (briefly, *Through the Looking Glass*). It is a sequel and a postscript in a similar way that *Time-Out of Time* is a postscript to the *Nuclear Time Travel.* It discloses Alice's fantastic world for a second time. Alice finds out that time is reversed, logic is reversed, and she can preserve her location just by running faster. She comes closer to an object if she moves away from it. This is an essential property of quantum motion. We do not have a resting coordinate system and no zero origin that we can refer to. But the ground state of motion is maximum unrest, an indefinable chaos, restless motion beyond definite dimensions. This is not to be confused with quantum fluctuations, because such fluctuations are already the outcome of quantum motion within stabilized material environments. We do not yet have sufficient tools to represent such motion. We can only see the results that seem to show us a stable world, our measurable matter. Even the present quantum physics is an outcome of some artificially filtered and stabilized quantum motion, a movement that is originally carried out in a restless disjointed space that is totally inaccessible to us, away from dimensions that are known to us.

"Above the door frame leading into the lab that contains the Looking Glass project at facility S4 at Area 51 there is a stuffed White Rabbit holding a backward watch mounted permanently to the frame" [54], Burisch remembered, and "Level 2 or '4-2' was known as '*Alice's Floor.*' This specific floor contained a laboratory for weapons research and development, three board rooms, and provisions for emergency supplies. Also located on level 4-2 were two specific areas which contained components for project sidekick. Level 4-2 was also the location of Project Looking Glass. This device *utilized six (composite) electromagnetic fields*, and a height adjustable rotating cylinder which is injected with a *specific type of gas.* The entire assembly can be rotated 90 degrees from the horizontal axis. This allows scientists to warp the local fabric of space-time both forward or backwards by long or short distances relative to the present time. The Project Looking Glass device was used to predict the potential probability of future events. Once the device is tuned properly, images of probable future events are projected in open space within the fields, similar to a hologram."

This gas was argon. It was used as an injection locking medium in a phase conjugating looking glass. That this device *utilized six (composite) electromagnetic fields* is simply because we have to take three areas and three perpendicular directions of propagation into account for the calibration. In short, the Euclidean space has three dimensions and the numbers 1, 2, and 3 allow for 6 permutations. We always come across these six degrees of freedom. In Minkowski space we have the six color spaces as represented by the hexagram.

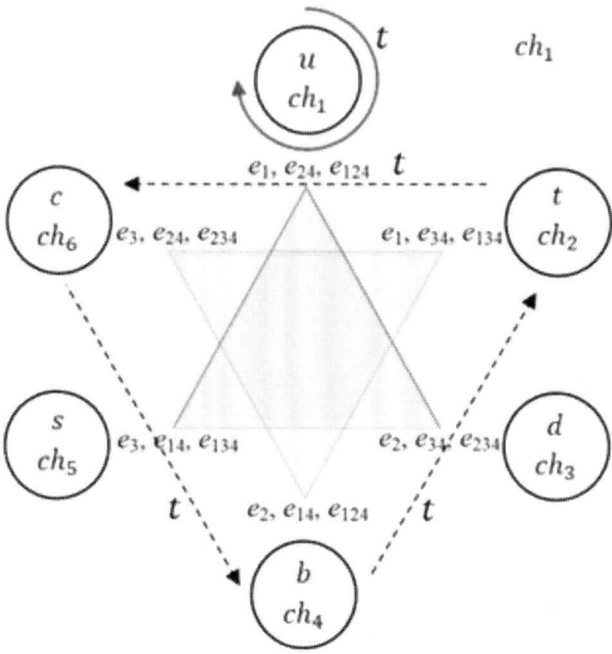

Figure 7. Six isomorphic maximum Cartan subalgebras in $Cl_{3,1}$.

In the 1970s six allrounders in laser research worked in Los Alamos National Laboratory. They were specialized in fields like laser fusion, plasma physics, quantum electronics, nonlinear optics, biophysics, molecular physics, infrared lasers, chemical lasers, gas lasers, degenerate four wave mixing, and optical phase conjugation. In 1982 they published their famous paper "*Through the Looking Glass* - with Phase Conjugation" [46]. This was actually late in relation to the early works of Bill Uhouse, but early for the topic of optical phase conjugation. They began by quoting Alice:

"'I don't understand ...' said Alice. 'It's dreadfully confusing!'

"'That's the effect of living backwards,' the Queen said kindly: 'it always makes one a little giddy at first–'

'Living backwards!' Alice repeated in great astonishment. 'I never heard of such a thing!'"

Lewis Carroll

A phase conjugator is a special type of mirror that reflects the light in the same direction from which it is incident. A conventional mirror reflects the light by inverting only one of the two normal components of the light beam. This leads to a change of direction. The phase conjugating mirror, however, inverts both components. As a result, the direction is maintained; there is essentially a reflected wave that follows the time-reverted path. Due to the interference of the incident and the phase conjugated outgoing wave, a correction of distortions and noise in the incident wave is possible, as if it had never occurred. With phase conjugation you can pull an image into a synchronous optical pattern so that you can edit it. You can enlarge the presence of optical phenomena.

The paper begins with an unorthodox statement: "Imagine a mirror that reflects more light than was incident, that reflects a beam into the same direction regardless of the mirror's tilt, that eliminates image distortions by causing light rays to retrace their paths as if running backward in time, and that when looked at allows the observer to see absolutely nothing. Science fiction, you say? Well, such mirrors have been the subject of intense investigation both here at Los Alamos and at other research laboratories around the world. Not only do they exist, but their practical applications may be far-reaching."

In ResearchGate you can find a group of authors whose work Burisch apparently quotes. They admit the following: "However, for performing the conjugation, the desired optical field should be exactly regenerated by the SLM, which requires extremely precise alignment and calibration between the camera and the SLM in six degrees of freedom. Therefore, ambient perturbations and noise can easily disable the functionality of

DOPC. This vulnerability has severely discouraged the practical use of DOPC" [47, p. 1].[16]

In the early devices, two looking glasses were used to receive sound images or acoustic messages from the past and future, but one encountered difficulties. In fact, this also has to do with the phase conjugate light reflection of laser light. These processes are not unknown today, but physicists still have insufficient control over them.

> "More direct demonstrations have also been performed using acoustic waves and microwaves, where conventional electronics can directly measure and generate the phase profile in real time" [47, p. 1].

Today's rather cautious to pessimistic-sounding assessments of the method of optical phase conjugation actually show clearly what happened at that time, not on a physical but on a sociological level. I assume that the former use of the looking glasses or the yellow book of the Orions only worked because the cooperation between earthly and extraterrestrial scientists was undisturbed. If it is true, as Burisch says, that MJ 12, the government and intelligence agencies, have decided to stop using the looking glass technology and the associated prediction of future events, then the matter apparently fell asleep. The scientists actually had a hard time understanding the theory behind these unusual methods, as I have shown giving the example of Mr. MacPherson in *Nuclear Time Travel* (NTT).

> "There is an UNKNOWN METHOD by which the atoms are changed. Instead of ionic or covalent bonds, this new system creates a THIRD TYPE of bonding principle. This new type of bonding causes atoms to take on a different form. Atoms go from a normal state – and when I say 'normal,' I mean normal in OUR science – to a 'multiplexed' form" [1, p. 40].

---

[16] DOCP = digital optical phase conjugation.

And further on:

> "Two plates of metal sit inside a chamber. This chamber is a vacuum. The chamber is a PERFECT VACUUM, that is, there is no outside air or other gases inside this chamber. A tube running from one chamber into another, pushes a gas inside. A third chamber forces still another gas to enter the first chamber. We know that the gases push against these two plates of metal and cause an ENORMOUS AMOUNT OF ENERGY TO BE CREATED.
>
> "Lastly, just before the gases are 'mixed,' there is a form of light that is focused against the outgoing energy. This light is of a lower frequency. Normally, in our quantum theory, a small frequency means a lower energy. However, is this system, the lower frequency of light was measured to contain an enormous amount of energy exerted upon this outgoing energy. When I left in 1988, we could NOT understand how the gases were mixed nor could we determine HOW the light was generated." [1, p. 51].

The following accident was reported by Michael Wolf and MacPherson: "In 1973, a terrible accident occurred at the Complex III facility. The propulsion system on the BC [comment: Blue Chariot] was activated and an explosion occurred killing six (6) technicians. Actually, it was NOT an explosion. A bolt or plasma ball came out of the propulsion system and struck the technicians, VAPORIZING THEM" [1, p. 48].

The question immediately arises, of course, of what the optics of the phase conjugate reflection have to do with such massive physical processes. What has nonlinear optics to do with materials science and high-energy physics? The answer is: The physics of the looking glass relates to the physics of extraterrestrial aircraft, just like optical phase conjugation relates to the optics of nuclear forces. For this reason, UFOs, in particular the flight simulator and the looking glass, as were familiar to Bill and Will Uhouse, belong together just as well as the looking glass and the physics of the stargate and the physics of the spaceships that can travel in hyperspace; the physics of all these gadgets is based on the controlled interference of phase conjugate reflections. You can easily see this with the help of King Solomon's Seal above. We know that there are no electrons in the electromagnetic waves of light, but light has an electrical component that interacts with electrons. Likewise, the light of the subnuclear fermions that constitute nucleons, but are not to be confused with the gluons, has no quarks in it directly, but it interacts with the quarks. Under the conditions we are most familiar with, the subnuclear light is reflected internally at the boundaries of the atomic nuclei. It does not get to the outside, but its reflection takes place in such a way that the association of protons and neutrons is stabilized. It is obvious that starships should not disintegrate, but rather they should remain stable under the most extreme conditions. The theory is that the phase conjugation of subnuclear light is such that the matter is stabilized and it is enclosed in a kind of gravitational vacuole.

Now I should explain to you what is meant by subnuclear light.

If you go through the representations in [48] you find out that the negative idempotents denote the energy density of anti-quarks. It is a remarkable feature of this system that an electromagnetic decomposition of the fermions is equivalent to a reconstruction of the corresponding anti-fermions. In addition, both the energy density and the wave-functions of the fermions are accessible to a phase conjugate reflection. The oscillating energy density is propagated by quasi plane waves in the quaternion color spaces of the following kind:

$$\varphi' = \exp\left(i\frac{p^{(1)}}{\hbar}\frac{1}{2}(Id - e_1 - e_{24} - e_{124})\right) = \left(\cos\frac{p^{(1)}}{\hbar}\right) Id + i\left(\sin\frac{p^{(1)}}{\hbar}\right) r_1 \text{ n} \quad (20)$$

where we take the first color space isomorphic with the fourfold ${}^4\mathbb{R} = \mathbb{R}\oplus\mathbb{R}\oplus\mathbb{R}\oplus\mathbb{R}$ as the propagated space.[17] The r_1 is the basic reflection of the first annihilating primitive idempotent, namely $r_1 = Id - 2f_{1,1} =$

[17] It is therefore that we can speak of quasi plane waves.

$\frac{1}{2}(Id - e_1 - e_{24} - e_{124})$. Note that r_1 behaves like the line element e_1 where norm is concerned, namely for both $e_1^2 = Id$ and $r_1^2 = Id$. Quantity $p^{(1)}$ is the impulse in "first color direction" r_1 and respectively e_1, quantity $\hbar$ denotes the reduced Planck constant. Now consider the factor of the space-time area component:

$$\mathrm{f} = exp\left(i\frac{p^{(1)}}{\hbar}\frac{1}{2}(-e_{24})\right) \tag{21}$$

The wedge product $-\frac{p^{(1)}}{2\hbar}e_2 \wedge e_4$ allows for two manifestations having equal direction of space-time area, through reverted both space and time arrow:

$$\mathrm{f}_a \stackrel{\text{def}}{=} \frac{p^{(1)}}{2\hbar}(-e_2 \wedge +e_4) \text{ and } \mathrm{f}_b \stackrel{\text{def}}{=} \frac{p^{(1)}}{2\hbar}(+e_2 \wedge -e_4) \tag{22}$$

Hence the only line element that is sensible for a phase conjugation is the propagation direction e_1 [49, p. 166]. Thus, the wave can be reflected with all its spatial properties by just reverting time. The energy can go forward by propagating backward. Finally, we may ask if that holds for the spinors too. Consider a spinor of, say, a red u-quark [49, chapter 12]. It has an area-extending component:

$$\varphi_r = \frac{1}{2}(e_1 + e_{24}) \text{ and torsion momentum } \iota_r = \frac{1}{\sqrt{2}}(e_3 + e_{13}) \tag{23}$$

The spinor multivector representative is essentially given by their Clifford product:

$|u\rangle = \frac{i}{\sqrt{2}}\varphi_r^{\wedge}\,\iota_r^{\wedge}$ with '^' indicating the main involution and $i = \sqrt{-1}$.

This is equal to magnitude:

$$|u\rangle = \frac{i}{4}(e_3 - e_{13} - e_{234} + j) \text{ with the 4-volume } j = e_{1234} \tag{24}$$

Now observe the following: Again, you can revert both e_2 and e_4 without changing the area_extending component φ_r. You can preserve that and conjugate phase by reverting direction e_3. Next, one can also preserve the direction of the spatial area e_{13} in the torsion momentum ι_r theoretically by reverting both directions e_1 and e_3 simultaneously, but that would imply that the first summand in the torsion $\iota_r = \frac{1}{\sqrt{2}}(e_3 + e_{13})$ is automatically reverted, which actually amounts to a flip of the spin. Note that the ι_r satisfies the familiar angular momentum algebra. Hence, if we finally flip the spin, we obtain a fermion spinor that represents motion with reverted direction e_1 only. The multivector history reads:[18]

$$|u\rangle = \frac{i}{4}(e_3 - e_{13} - e_{234} + j) \text{ flip } e_4 \text{ and get} \tag{25}$$

$$\Rightarrow \frac{i}{4}(e_3 - e_{13} + e_{234} - j) \text{ flip } e_3 \text{ and get} \tag{26}$$
$$\Rightarrow \frac{i}{4}(-e_3 + e_{13} - e_{234} + j)$$

This is the wayback fermion with the same torsion. Bottom line is that in the atomic nuclei the nucleons are stabilized by phase conjugated waves of subnuclear light. The looking glass technology of Alice would therefore be very well applicable in nuclear physics. Of course, the differences in force compared to non-linear photonics are enormous and the technical difficulties cannot be overestimated. I think that is exactly the crux of the matter – why earthly technicians didn't quite understand the alien spaceships and a few important things went terribly wrong. We have not been able to produce hyperspace traveling craft. Maybe a few

[18] I explained backward propagation of subnuclear energy at the 12th International Conference on Clifford Algebras in Henei, China, http://smartchair.org/hp/ICCA2020. The manuscript can be found here: https://www.researchgate.net/publication/346726171_A_Cultural_Timeline_of_Relativistic_Logic_Quaternions_and_Phase_Conjugation.

small groups that could afford it have broken away from space companies and NASA and are already floating around elsewhere. But global civil space travel and science are not affected. Few know the physics of antigravity. There are many theories but only a few have correctly described the drive. Bob Lazar is one of them. He said that by giving an angular momentum to moscovium (formerly called ununpentium), because of its symmetry the strong force exceeds the limits of the atomic nucleus. No doubt he understood some of the process and described it correctly. But nobody believed him. Maybe the initiates believed him, but they wanted to take advantage of this knowledge alone, and because as a young man he was thrilled and not quite "tight," they excluded him.

Chapter 5

HISTORY AND FUTURE DEVELOPMENT OF GRAVITATIONAL PROPULSION

The history of gravitational propulsion has two branches. It bifurcated more than 60 years ago into a civilian and a secret military threat. After some more or less fortunate landings of extraterrestrial spacecraft, the US military began to reconstruct a few drives that apparently controlled gravity. At the end of the 1950s, after the Kingman accident (1953), civil research followed suit and devoted itself to the topic. The first relevant contributions on the subject of controlled gravitational wave radiation came from Leopold Halpern [50]. He was assistant to Erwin Schrödinger from 1957 to 1959 and did research on gravitation at CERN and the Niels Bohr Institute in Copenhagen. From 1974 to 1984 he worked as a Senior Research Associate with Paul Dirac. Some physicists have followed Halpern's idea, for example Baker [51], Fontana [52], Alekseev [53], and proposed a series of devices and generators taking the form of microelectromechanical systems (MEMS generator), high-temperature superconductors (HTSC) and GASERs (gravitation amplification by stimulated emission of radiation), some gravitational counterpart of the LASER in theory. In 2012, Giorgio Fontana published an article about the HTSC GASER [54] on ResearchGate. At the same time, a trend-setting paper was published by

STAIF [55]. We are, of course, right to ask ourselves why it has not been possible to produce a functioning device in the past twenty years? The theoretical workload and preparations for construction plans are time-consuming. Were all these efforts in vain? It cannot be assumed that useful research reports in the civilian sector will not be published at least in part. So, what happened? What situation are we actually in? To answer that question, let me go back to the beginnings of the Copenhagen School, because Leopold Halpern followed this school.

Let us remember the illustration of quantum physical events, say observation of quanta in the early days. Imagine a screen behind a double slit, or some kind of imaging device that shows incoming particles in a wire chamber, or particle collisions in a bubble chamber. We don't see any waves there, but, first of all, we observe the flashing of something that we call quanta. Events become visible. Today we say that the wave function collapses there. But which wave? Well, if we collect these events that appear in some region with more or less probability, if we collect these over a period of time, then we get a synchronous picture of events with varying density. The picture in the here and now resembles the plot of a wave. So, we suspect something like a guiding wave that propels some particles. In any case, the original idea that the dynamics of quanta represents a statistical ensemble and that the wave function possibly says something about the probability of occurrence of events was a very good idea.

If you look up in Wikipedia today what the Copenhagen interpretation is, and if you translate the German entry into English with Google, you will get what the author had learned from Walter Thirring in Vienna. This equals what Leopold Halpern knew too:

> "The Copenhagen interpretation, also called the Copenhagen interpretation [sic], is an interpretation of quantum mechanics. It was formulated around 1927 by Niels Bohr and Werner Heisenberg during their collaboration in Copenhagen and *is based on the Born probability interpretation of the wave function* proposed by Max Born. Strictly speaking, it is a collective term for similar interpretations that have been differentiated over the years. The version, which is also referred to as

> the standard interpretation, is based in particular on John von Neumann and Paul Dirac." [2][19]

This I regard as obliging. However, the English Wikipedia entry does not include the "Born probability interpretation" in the sentence, but reads as follows:

> "The Copenhagen interpretation is a collection of views about the meaning of quantum mechanics principally attributed to Niels Bohr and Werner Heisenberg. It is one of the oldest of numerous proposed *interpretations of quantum mechanics*, as features of it date to the development of quantum mechanics during 1925–1927, and it remains one of the most commonly taught. There is no definitive historical statement of what is the Copenhagen interpretation. There are some fundamental agreements and disagreements between the views of Bohr and Heisenberg."

These statements seem clear and they are sociologically relevant. However, they are not only of sociological importance, but relate to the content of a phenomenology of gravitational waves. In physics, in our bubble chambers and elsewhere, we had those probabilistic events, quanta that seemed to emerge out of nowhere, or out of some wave background, if you will. In general relativity we did not have such events, no quantum, no graviton, and there was no wave guiding quanta. We rather speculated about a quantum of spin-2 quadrupole radiation. Is there anything substantial that general relativity and quantum theory did have

---

[19] Die Kopenhagener Deutung, auch Kopenhagener Interpretation genannt, ist eine Interpretation der Quantenmechanik. Sie wurde um 1927 von Niels Bohr und Werner Heisenberg während ihrer Zusammenarbeit in Kopenhagen formuliert und basiert auf der von Max Born vorgeschlagenen bornschen Wahrscheinlichkeitsinterpretation der Wellenfunktion. Es handelt sich genau genommen um einen Sammelbegriff ähnlicher Interpretationen, die mit den Jahren ausdifferenziert wurden. Besonders auf John von Neumann und Paul Dirac fußt die Version, die auch als Standardinterpretation bezeichnet wird [2]. Seite "Kopenhagener Deutung." In: Wikipedia, Die freie Enzyklopädie. Editing status: 11. November 2020, 07:58 UTC. URL: https://de.wikipedia.org/w/index.php?title=Kopenhagener_Deutung&oldid=205405459 (accessed February 4, 2021, 21:09 UTC).

in common in those days? I would say there were two essential commonalities:

- equations of motion within a classic space-time frame,
- Lagrange functions for such equations.

Gravitational waves (GWs) were first proposed in 1905 by Henry Poincare and predicted by Albert Einstein in 1916. However, a direct observation of GWs was not made before 2015, namely by the LIGO gravitational wave detectors. Gravitons did not yet appear on any screen.

But, it is interesting that Einstein had concluded in those days that "a perfected quantum theory would have to be extended also to the gravitational theory in order to give an unambiguous description of the situation."[20] So Einstein saw the necessity of beginning at the foundations of quantum theory and he saw the need to work his way up to the general theory of relativity. If the later results of high energy physics had already been known at that time, then I think he would have seen the possibility of starting from the symmetries of space-time algebra and then recognizing the symmetry of the elementary particles. At that time, however, the incompleteness did not primarily concern the physically available experiments, but there was and is still incompleteness in the theoretical area. This is only eliminated when the very special meaning of the Clifford algebra or, more generally, of the geometric algebra for physical spaces is recognized.

- We can build a stable bridge between the theory of relativity and quantum mechanics by not primarily explaining the curvature of space through matter, but by tracing the familiar quantum numbers back to the peculiarities of the geometric Minkowski space-time algebra.

---

[20] Quoted after Halpern and Laurent 1964 [50]: Einstein [56, p.154], [57, p. 688].

It seems to be comparatively easy to design a Lagrange function in correspondence with Einstein's or Papapetrou's equations of motion. But, as Halpern figured out, "it is well known that these equations actually are not consistent, but we hope that they may nevertheless serve to evaluate the gravitational field approximately. We in the following consider time-periodic solutions of the linearized equations." [50, p. 730]. Halpern linearized the Lagrange function and examined its zero-th approximation. His theory is similar to that of Gunnar Nordström's theory of gravitation in flat space, which also ended up with a Klein Gordon equation of motion. Hence, he quantized a gravitational field in flat space and compared possible gravitational and electromagnetic transitions in some hypothetical oscillating field, the existence of which was and still is not at all certain. But now it was theoretically possible to apply the Clebsch-Gordon formalism in the customary way to calculate orbital angular momentum eigenfunctions and investigate and compare creation and annihilation of electromagnetic and gravitational flux; spherical Bessel and Hankel functions took the stage, and the intensity of quadrupole radiation could be estimated. The fact is, however, that the supposed gravitational quadrupole radiation in the far infrared has not yet been observed, and that is probably not due to the low intensity of the radiation alone.

In order to be more successful in this technical matter, I have suggested two things. First, we have to consider the creation and annihilation operators of actually existing vibrational fields, because it should not happen that the existence of quantized fields can simply be neglected because of inadequacies in the mathematical approach. Second, we have to take into account that the energy of physical processes is not primarily observed in a space-time frame, but rather in that Clifford space that is generated by space-time. I therefore summarize:

- Gravitation is a peculiar appearance of the strong force. It is not an "ultraweak force."
- The empirically verified oscillators of the field are given by isospin, color, and flavor rotations.

- The macroscopic observation space of space-time is the Clifford algebra that is generated by the Minkowski space in the opposite (Lorentz) metric.
- Theory should predict the familiar quantum numbers like isospin, hyperspin, charge, color, flavor, and mass.

For the past 20 years I have tried to combine the empirically observable reality of high-energy physics with the mathematical complexity, richness of representation, and elegance of Clifford algebra. It occurred to me that the use of special mathematical tools in this case would have enormous consequences. It is clear that the 16-dimensional Clifford space generated by Minkowski space, like any vector space, contains a set of linearly independent basis vectors. This independence is substantial. It could be the case that:

- Energy needed to be coordinated in a completely new way.

Interestingly, in his 2012-paper, "listing new directions," Fontana brings in the "Dineutron Upconverting Transducer" which is based on a nuclear model proposed by Linus Pauling. This model seems important as it explicitly emphasizes the isospin of nuclear particles. But any possible radiation of neutron-tunneling through alpha-particles is not calculated there.

In my work it soon became apparent how important a role the isospin, the geometry of the symmetric unitary group $SU(3,\mathbb{C})$, and its non-compact counterpart $SL(3,\mathbb{R})$ would play for both space-time and particles. The choice of the metric for the Minkowski space turned out to be by no means insignificant.

I recall, in October 2004, that Fontana [58], following an analysis of the "(gravitational) instanton," had argued that it "can be defined as a pseudoparticle solution of the Einstein equations in Euclidean space-time (signature $++++$) with a cosmological constant. The Euclidean signature helps to remove singularities found in the theory with

Lorentzian signature $(-+++)$ and has been introduced for the purpose of the analysis."

This holds true if we concentrate on curvature and removal of singularities. But as soon as we focus on quantum mechanics and especially high-energy physics, the Clifford algebra generated by the Minkowski space in the Lorenz metric proves to be particularly advantageous. Here, I would like to quote a word from our late friend Pertti Lounesto [4]. He spoke of Minkowski algebra "tilt to the opposite metric." Some physicists turn from Clifford algebra $Cl_{1,3}$ to its opposite algebra $Cl_{3,1}$, replacing the gamma-matrices γ_μ by their product $i\gamma_\mu$ with the imaginary unit. Thus, they obtain a multivector-space that is isomorphic with the Majorana algebra of $4 \times 4 -$matrices with real valued entries. Lounesto denoted this as $Mat(4, \mathbb{R})$. I always use his notation. This "opposite" Minkowski algebra contains lattices of idempotents that correspond with the subspaces of colored fermion states with flavor, hyperspin, charge, and all the rest of it. In a very extensive research work that extended over many years, I worked out some necessary fundamentals that concern the meaning of the Clifford monomials, of the primitive idempotents and the associated basic reflections, as well as some necessary, new involutions that were denoted as transpositions. At last, I received an illuminating partition of the whole algebra into three parts, one with positive signature and dimension 10 and two quaternion spaces.

The diagram of thermodynamic trigonal rotations as depicted on the upper right was discussed in *Nuclear Time Travel and the Alien Mind* [1, pp. 80ff]. It means that isospin, color, and flavor rotations imply corresponding rotations in the thermodynamic space of the triple $\{e_4, \mathcal{V}, \mathbb{V}\}$ with unit 3-volume $\mathcal{V} = e_{123}$ and space-time volume $\mathbb{V} = e_{1234}$. The strong force in its highest symmetry is flavor-color-time-space locked. Now that we have the equations of the thermodynamic oscillator [1, equ. 21, 25], we can quantize it, given the amplitudes of the oscillating enclosure volumes. The quantized 4-volume is a direct consequence of the strong interaction. Mathematically it causes the confinement of fermions. The quantization procedure is straightforward. Its empirical realization, however, is still a long way off.

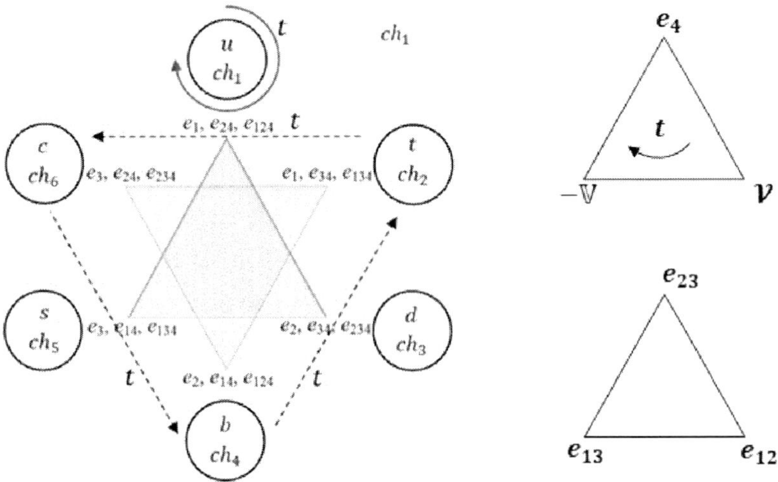

Figure 8. Ternary strong force-partition of the Minkowski algebra.

But to get that far at all, it took a lot of thought and patience. At the beginning of the 1990s I had already realized that the categorical features of the known interaction forces could be derived solely from the symmetry properties of space and time that we recognized in the geometric algebraic structures. I then published *The generative process of space-time and strong interaction* with subtitle "quantum numbers of orientation" [59], and, of course, orientation now meant much more than PCT-(in)variance. This is because the coordinate automorphisms of the Clifford algebra had a much higher complexity. By now it had become clear that the mathematical instruments had to be examined in more detail with regard to their usefulness for physics. There were several Clifford algebras, their complexified versions, matrix algebras over different fields, and the Dirac algebra, which could be used to represent quantum physical and space-time processes. The mathematics of the curvature of space was not an ostensible concern, because the curvature tensor we needed would result from the Lie algebras anyway, once these were found. I wrote, then, in the abstract to *Minimal Spin Gauge Theory* [60, p.

63], that "the symmetric unitary group $SU(3,\mathbb{C})$ is not a part of any real Clifford algebra of 4-dimensional space, especially not of the algebra $Cl_{1,3}$ of the Minkowski space-time, nor of the algebra $Cl_{3,1}$ in the opposite metric." In the beginning of my rather extended research there were many who made me aware by asking me, "but what does $SU(3,\mathbb{C})$ have to do with Clifford algebra?" The first of these was Johan Gijsbertus F. Belinfante, when I met him sometime in the late 1990s. Things were not yet well known, then, but I was determined to hold on to it and move on. Zbigniew Oziewicz, Rafal Ablamowicz, José Vargas, and Bertfried Fauser supported me, as they saw the relevance of the idea. Oziewicz invited me to Mexico, and Fauser and Dehnen invited me to Konstanz to talk it over. So, I went on asking how quantum chromodynamics enters into the geometry. A first answer was that "the group $SU(3,\mathbb{C})$ is an object of both the complexified algebras $\mathbb{C}\otimes Cl_{1,3}$ and $\mathbb{C}\otimes Cl_{3,1}$. To show this we first define six color spaces which are spanned by conjugate triples of commuting base elements. These contain the six idempotent lattices that can be located in $\mathbb{C}\otimes Cl_{3,1}$. Their images exist in both $\mathbb{C}\otimes Cl_{1,3}$ and $\mathbb{C}\otimes Cl_{3,1}$. Further in each color space there is defined an octahedral orientation stabilizer group which fixates one lepton and color rotates the states in its quark family" [60, p. 63].

Most important were the relationships:

$$Cl_{1,3} \simeq Cl^{0}_{1,4} \simeq Cl^{0}_{4,1} \subset Cl_{4,1} \simeq Mat(4,\mathbb{C}) \tag{27}$$

These equations demonstrate the outstanding importance of the Dirac algebra, and at the same time legitimate the use of the Majorana algebra that contained the strong force lattices in a most natural way, namely we have:

$$Cl_{1,3} \not\simeq Mat(4,\mathbb{C})$$

but

$$Cl_{1,3} \subset Mat(4,\mathbb{C}) \tag{28}$$

whereas

$$\mathbb{C}\otimes Cl_{3,1} \simeq Mat(4,\mathbb{C}) \tag{29}$$

Those who really want to understand the theory, and you will soon understand that we are not dealing with a model here but with a new theory, I strongly recommend going into the above paper [60] as it discusses the first perceptions that are new in the foundations of physics. In it are explained the six basic color spaces, the minimal representations of their generating primitive idempotents, the basic reflections that create the discrete automorphism groups, the Gell-Mann matrices, exemplified by the first of the six color spaces, and a representation for a physical Clifford algebra having an imaginary unit that commutes with the whole algebra, that is, $Cl_{1,7}$. I did this out of gratitude and in response to Roy Chisholm. We always agreed on one point of view: Chisholm and Farwell had argued that "*there is no distinction between space-time and the internal interaction space,*" which I have also believed since my earliest days in the Schrödinger-Zimmer (Schrödinger Room).

SPACE ELEMENTS

It was after I had written the above paper [60] that the extraordinary importance of a 6-partition of the Clifford algebra $Cl_{3,1}$ into commutative modules with 1-norm became visible, and I mentioned that in a lecture titled *Extension theory and Clifford algebra with transposition* at the 6th Conference on Clifford Algebras (2002).[21] In the corresponding book chapter *Transpositions in Clifford Algebra* [61, p. 359], is the decomposition of the graded color spaces into fourfold real fields:

$$Ch_{\chi} \simeq \mathbb{R}\oplus\mathbb{R}\oplus\mathbb{R}\oplus\mathbb{R} = {}^{4}\mathbb{R} \tag{30}$$

[21] Program of the 6th Conference on Clifford Algebras, Friday, May 24, 2002 (tntech.edu).

where zero-valued 1-norm was explained for the first time using minimal isospin-ideals in an exact way. In the 4-fold real ring ${}^4\mathbb{R}$ the basis of ch_1 can be represented by the quadruple numbers:

$$Id = [1,1,1,1];\ e_1 = [1,1,-1,-1];\ e_{24} = [1,-1,1,-1];\ e_{124} = [1,-1,-1,1] \quad (31)$$

Apart from the identity Id these are unit line, and area and volume elements with a zero sum of entries. Since the discovery of their geometric importance for quantum physics, I have spoken of quantum motion rather than of quantum mechanics, and we should also consider "unit space elements" independent of their grade. In its greatest transcendental extension, quantum motion is a zero-sum game. It knows nothing of the "time" perceived by us humans, and, as a global and local system, it also knows no conservation of energy and only very conditionally the conservation of angular momentum. The unit elements into which primitive idempotents are decomposed have grades 0, 1, 2, and 3 as you can see. The wonderful thing about these magnitudes is that they can be interpreted as energy carriers with the 1-norm energy zero. Their superposition is not zero, but is capable of locating the unit energy density at the primitive idempotents that represent pure fermion states.

The New Interpretation of Quantum Motion

It is a mistake to assume that the Copenhagen interpretation is irreconcilable with David Bohm's interpretation. In addition, quantum mechanics is not mechanics, but a biodynamic process, which is why I decided to speak about quantum motion two decades ago. The pictures of Bohr, Heisenberg, and others are brought forth by transdimensional energy processing. This I denoted as *quantum motion*. It is that third dynamics of indefinite dimensional events which Bohm said are neither wave nor particle. They bring about measurable space-time as we know it at present, but neither move in any space with definite dimension nor in time. Quantum motion happens apart from time. This idea was published in *On Motion* [62].

Our current knowledge of quantum motion and entanglement is rather limited. I have therefore always endeavored to infer the transdimensional movement from the events and from procedural manifestations on the macro level. I have been investigating special phenomena of entanglement in [5, pp. 126ff, 177ff], the sections on universal gates and two-qubits in Clifford algebra, but what is even more important concerns vacuum-conductivity and …

COLOR-FLAVOR-TIME-SPACE LOCKING

Even if we ignore the strong force and only deal with electrical conductivity of the vacuum, we come across fundamental questions, which, among other things, concern the so-called natural constants. There is some confusion here. In an important article, Friebe [63] pointed out how Maxwell defined the dielectric displacement and the magnetic induction in a vacuum. He assumed them to be constant regardless of time, and finally the speed of light had to be derived from an equation by definition:

$$\varepsilon_0\mu_0 = 1/c^2 \tag{32}$$

In particular Maxwell proposed fixations:

$$\varepsilon_0 = 1 \; \mu_0 = 1/c^2 \tag{33}$$

and alternatively:

$$\mu_0 = 1 \;\; \varepsilon_0 = 1/c^2 \tag{34}$$

Heaviside and Hertz, however made use of:

$$\mu_0 = 1/c \quad \varepsilon_0 = 1/c \tag{35}$$

Such assignments gave Maxwell's equations a symmetrical form. Friebe then decided as follows:

> "In the case of systems moving towards one another, it is not space and/or time that are to be relativized, but only the permeability μ_0 of the vacuum, as a variable that depends on the relative speed, has to be taken into account (Friebe 1980, p. 15). Because μ_0 characterizes a magnetic effect, which is the result of moving electrical charges. In this sense, μ_0 is only a speed conversion factor (ε_0 remains a constant given by the measurement system)"

The ambiguities go even further, because we can ask what is the magnitude of the vacuum conductivity. Many claim that the vacuum is a perfect insulator. I don't list up anyone who still thinks so. Few, however, realize that it is rather a perfect conductor. We have to differentiate between the properties of the vacuum and the properties of the electrodes that emit particles into the void. This is not primarily about the so-called Fermi level, which is another important story. Among the few who are attentive to real events is Charles Chandler [64, p. 1], who observed: "The resistance of a vacuum is commonly quoted as being infinite. Here's how Wikipedians put it:" So they start out saying that a perfect vacuum is a perfect insulator. Then they say that heat and/or electric field at the electrodes can liberate electrons, and then devices can utilize the resulting vacuum conductivity. Due to the legacy of shifts in EM theory as it evolved, such self-contradicting 'explanations' are not uncommon. But the development of any fully mechanistic model incorporating EM requires an accurate definition of what EM actually is. The resistance of a vacuum has nothing to do with the electrode's work function, nor its ionization potential. The resistance of a vacuum, if it exists, is a property of the vacuum itself, and needs to be measured independent of the properties of the electrodes." It was at about the same time when I wrote

Minimal Spin Gauge Theory that another important paper was published by Rajagopal and Wilczek [65] in the Physical Review Letters demonstrating how the color-flavor locked phase enforces electrical neutrality. I discovered that paper only recently, and as a matter of fact the theory proposed by me in 2001 [60, p. 70] and calculations in *Algebra of Matter* [66, p. 280] produced the result that a pure space-time gauge of QCD in $\mathbb{C}\otimes Cl_{3,1}$ is rigorously connected with a color-flavor locking in the whole space. Rajagopal and Wilczek pointed out in their abstract: "We demonstrate that quark matter in the color-flavor locked phase of QCD is rigorously electrically neutral, despite the unequal quark masses, and even in the presence of an electron chemical potential. As long as the strange quark mass and the electron chemical potential do not preclude the color-flavor locked phase, quark matter is automatically neutral. No electrons are required and none are admitted."

The void behaves passively. It does not act on the movement of charged electrons. Likewise, it does not exert any force on gluons. Subnuclear fermions are coupled to each other by gluons and even to outer nuclear motion, but the vacuum does not act on gluon conductivity in matter. To approach gluon conductivity with the aid of the space-time group (Appendix 1), we have to realize the appropriate $SL(3,\mathbb{R}) \subset Cl_{3,1}$ and, respectively, the compact $SU(3,\mathbb{C}) \subset \mathbb{C}\otimes Cl_{3,1}$ to construct spinors.

Recalling *Nuclear Time Travel* [1, pp. 70ff]:[22] "Strong interactions come up in the innermost of the generative machinery of space-time. Let me highlight some important events. In the utmost corner of the seal there resides, if our system is properly tuned, a red u-quark. It is then represented by the primitive idempotent

$$f_{1,2} = \tfrac{1}{2}(Id + e_1)\tfrac{1}{2}(Id - e_{24}) \qquad \text{(NTT 17)}$$

Now consider, on the first color space ch_1 there may act a series of hexagonal, trigonal and other crystallographic operators. One of these

[22] The equations indicated NTT 17 to 21 are part of the citation from *Nuclear Time Travel* (NTT).

is $\boldsymbol{t} := R_\chi$ a *period 3 element* of the algebra capable to rotate color. It has the typical form of a rotation operator for primitive idempotents. Namely, it can be represented, as is usual for rotations, by two reflections. Those two are …

$$\begin{aligned} s_{2,3} &\stackrel{\text{def}}{=} Id - 2f_{2,3} \\ s_{6,2} &\stackrel{\text{def}}{=} Id - 2f_{6,2} \end{aligned} \qquad \text{(NTT 18)}$$

Their Clifford product brings forth the rotator $R_\chi = s_{6,2}\, s_{2,3}$ and its inverse $R_\chi^{-1} = s_{2,3}\, s_{6,2}$. We observe the following 3-cycle of color transformations within the first color space and ε some real scale factor

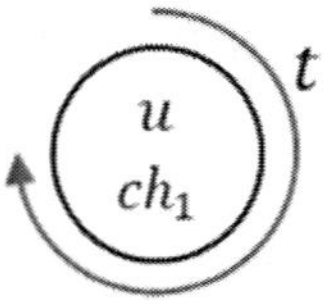

u
ch_1
t

$$\begin{aligned} R_\chi^{-1}\, \varepsilon f_{1,2}\, R_\chi &= \varepsilon f_{1,4} \\ R_\chi^{-1}\, \varepsilon f_{1,4}\, R_\chi &= \varepsilon f_{1,3} \\ R_\chi^{-1}\, \varepsilon f_{1,3}\, R_\chi &= \varepsilon f_{1,2} \end{aligned} \qquad \text{(NTT 19)}$$

with the first primitive idempotent $f_{1,1}$ in color space ch_1 preserved by R_χ. These equations signify the inner transformations triggered by $\boldsymbol{t}$, which acts on ch_1. There is trigonal action in ch_3 too:

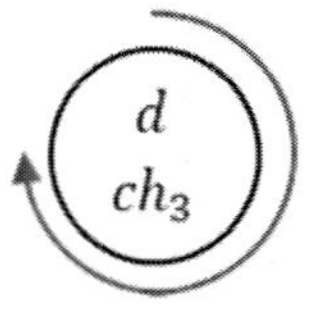

d
ch_3

$$\begin{aligned} R_\chi^{-1}\, \varepsilon f_{3,2}\, R_\chi &= \varepsilon f_{3,3} \\ R_\chi^{-1}\, \varepsilon f_{3,3}\, R_\chi &= \varepsilon f_{3,1} \\ R_\chi^{-1}\, \varepsilon f_{3,1}\, R_\chi &= \varepsilon f_{3,2} \end{aligned}$$

with the fourth primitive idempotent $f_{3,4}$ in color space ch_3 preserved. There is further action in ch_5:

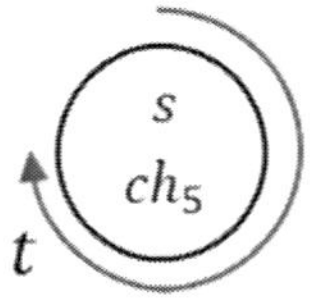

s
ch_5
t

$$R_\chi^{-1}\,\varepsilon f_{5,1}\,R_\chi = \varepsilon f_{5,4}$$
$$R_\chi^{-1}\,\varepsilon f_{5,4}\,R_\chi = \varepsilon f_{5,2}$$
$$R_\chi^{-1}\,\varepsilon f_{5,2}\,R_\chi = \varepsilon f_{5,1}$$

with the third primitive idempotent $f_{5,3}$ in color space ch_5 preserved by R_χ.

But, in addition to this ternary interaction, color is locked with flavor. That is the good thing about $\boldsymbol{t}$ – it preserves color spaces ch_1, ch_3, ch_5. It seems to act as a universal trigonal rotator. It does not send any element of ch_1 out of ch_1. Further it does not send any element of ch_3 out of ch_3, and it does not send any element of ch_5 elsewhere. But $\boldsymbol{t}$ does not preserve any elements within spaces ch_2, ch_4, ch_6. Rather, it maps those elements elsewhere. It sends color space ch_2 to ch_6, ch_6 to ch_4, and ch_4 back to ch_2 in a 3-cycle. Bottom line: The chaotic interplay of partons in nucleons follows certain rules which support macroscopic geometry. Note the orbit of $f_{6,2}$! No elements of ch_2, ch_4, ch_6 are preserved by $\boldsymbol{t} = R_\chi$:

$$R_\chi^{-1}\,\varepsilon f_{2,3}\,R_\chi = \varepsilon f_{6,2} \qquad \text{(NTT 20)}$$
$$R_\chi^{-1}\,\varepsilon f_{6,2}\,R_\chi = \varepsilon f_{4,1}$$
$$R_\chi^{-1}\,\varepsilon f_{4,1}\,R_\chi = \varepsilon f_{2,3}$$

But $\boldsymbol{t}$ carries out trigonal mappings of primitive idempotents from ch_2, ch_4, ch_6. These are flavor rotations.

We can regard each color space as some inner space for each flavor. The inner space has in it those color degrees of freedom that locally preserve a field's flavor. Yet the same force that may excite inner motion of color in one triangle, namely in the upright triangle of the David star, will trigger a movement of flavor in the inverted triangle of flavors.

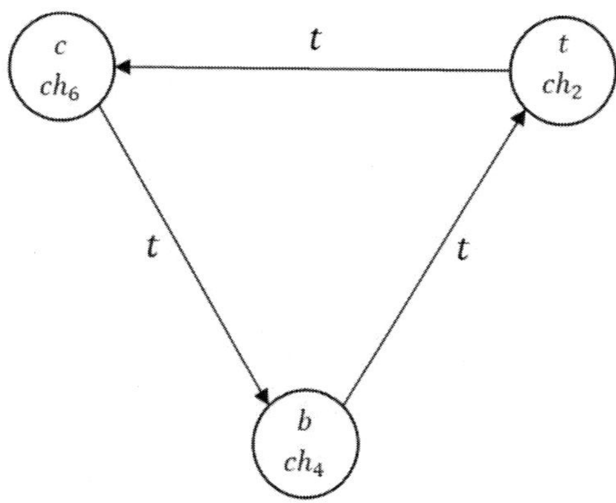

Figure 9. Trigonal mapping of color-spacing.

Note that the operation $\boldsymbol{t} = R_{\chi}$ mediates between what we tend to observe in physics as pure states. It connects fermions. Its purpose is not primarily to create superpositions of fermions. We shall come to that. But first let us realize what happens to the thermodynamic variables, that is, volume, time and space-time-volume. Motion in Ch is accompanied by analogous motion in the bivector-part and in the thermodynamic quaternion part of the system. A trigonal color and flavor rotation of fermions will also trigger a period 3 rotation in the thermodynamic time-space $\mathbb{T}$. With that special rotator we shall observe 'pure' rotations in the basis. I call it the *wheel of time.*

$$R_{\chi}^{-1}\, e_4\, R_{\chi} = e_{123} \qquad \text{(NTT 21)}$$
$$R_{\chi}^{-1}\, e_{123}\, R_{\chi} = -J$$
$$R_{\chi}^{-1}\, (-J)\, R_{\chi} = e_4$$

This may surprise us. But it is a fact which follows from the natural design of quantum space-time." Trigonal color rotations are locked with rotations of time-space. We realize there is still another way to calibrate the system. We can regard any single corner of the David star as a container of different colors rather than of different flavors. With that in mind, we may rename colorspace as flavorspace. Still, the trigonal and hexagonal rotations of fermion states will be locked with a trigonal or hexagonal rotation of the thermodynamic time-space.

Hence, it is now recommended to use the other structural calibration and regard corners in the hexagram as containers of distinct fermions, and rotate corners in order to represent color rotations.[23] Now it is interesting to note that the space-time groups $SU(3,\mathbb{C})$ that have been partly *discovered* as geometric features, partly *constructed* because I regarded that as important, do in fact have a form that is well known in the astrophysics of neutron stars. This is sometimes denoted as $SU_{C,F}(3,\mathbb{C})$ expressing thereby that color C is entirely locked with flavor F. The most peculiar finding is, however, that the vacuum CFL is additionally locked with the thermodynamic time variables. Thus, we have a universal CFTSL, a color-flavor-time-space lock for the void gluon ground state. When this lock is opened, the oscillator of the thermodynamic time-space discloses its many degrees of freedom, that is, its high degeneracy, which is driven by the QCD interactions. To calculate gluon spinors in this representation, we could perceive the Seal as depicted below

[23] There exist various degrees of freedom in the construction of symmetry groups with a pure space-time gauge, and beyond that we have many parameters that characterize the interaction forces. All these taken together constitute the material calibration of space-time.

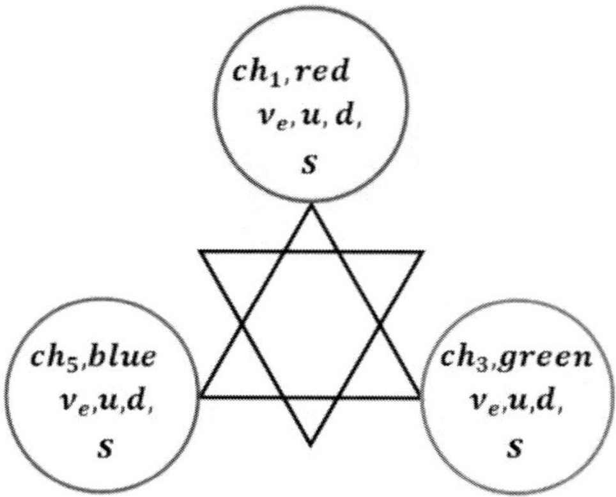

Figure 10. Ternary color partition of Minkowski algebra.

We proceed with the color spinors. Colored subspaces are commutative modules defined by the following quadruples:

ch_1 ... $\{Id, e_1, e_{24}, e_{124}\}$ ... red
ch_3 ... $\{Id, e_2, e_{34}, e_{234}\}$ ... green
ch_5 ... $\{Id, e_3, e_{14}, e_{134}\}$ ... blue

Red area extender torsion momentum symbolic operator ket[24]

red u, d, s

$$\varphi_r = \frac{1}{2}(e_1 + e_{24}) \quad \iota_r = \frac{1}{\sqrt{2}}(e_3 + e_{13})$$

$$|u_r\rangle = \frac{i}{\sqrt{2}}\varphi_r \hat{\iota_r} = \frac{i}{4}(e_3 - e_{13} + e_{234} - j)$$

$$|d_r\rangle = \frac{i}{\sqrt{2}}\hat{\varphi_r}\hat{\iota_r} = \frac{i}{4}(-e_3 + e_{13} + e_{234} - j)$$

$$|s_r\rangle = \frac{i}{\sqrt{2}}\hat{\varphi_r}\iota_r = \frac{i}{4}(-e_3 - e_{13} - e_{234} - j) \quad (36)$$

---

[24] Symbolic operator ket is a convention of notation, not a ket vector in Hilbert space.

Fix point of stabilizer $SU_{FC}(3,\mathbb{C})_{red}$ is

$$|\nu_{\mathrm{e}}\text{-}\rangle = \frac{i}{\sqrt{2}}\varphi_r\iota_r = \frac{i}{4}(+e_3 + e_{13} - e_{234} - j)$$

These should give us the correct energy densities. We calculate operator bra $\langle u_r| = \hat{u_r}$ and so forth and obtain:

$$\begin{aligned}
|u_r\rangle\langle u_r| &= \tfrac{1}{4}(Id - e_1 - e_{24} + e_{124}) = f_{1,2}\\
|d_r\rangle\langle d_r| &= \tfrac{1}{4}(Id - e_1 + e_{24} - e_{124}) = f_{1,3}\\
|s_r\rangle\langle s_r| &= \tfrac{1}{4}(Id + e_1 - e_{24} - e_{124}) = f_{1,4}\\
|\nu_{\mathrm{e}}\text{-}\rangle\langle \nu_{\mathrm{e}}\text{-}| &= \tfrac{1}{4}(Id + e_1 + e_{24} + e_{124}) = f_{1,1}
\end{aligned} \tag{37}$$

green area extender torsion momentum symbolic operator ket

green u, d, s

$$\begin{aligned}
\varphi_{gr} &= \tfrac{1}{2}(e_2 + e_{34}) \quad \iota_{gr} = \tfrac{1}{\sqrt{2}}(-e_1 + e_{12})\\
|u_{gr}\rangle &= \tfrac{i}{\sqrt{2}}\varphi_{gr}\hat{\iota}_{gr} = \tfrac{i}{4}(-e_1 - e_{12} - e_{134} + j)\\
|d_{gr}\rangle &= \tfrac{i}{\sqrt{2}}\hat{\varphi}_{gr}\hat{\iota}_{gr} = \tfrac{i}{4}(e_1 + e_{12} + e_{134} + j)\\
|s_{gr}\rangle &= \frac{i}{\sqrt{2}}\hat{\varphi_r}\iota_r = \frac{i}{4}(e_1 - e_{12} - e_{134} + j)
\end{aligned} \tag{38}$$

Fix point of stabilizer $SU_{FC}(3,\mathbb{C})_{green}$ is

$$|\nu_{\mathrm{e}}\text{-}\rangle = \frac{i}{\sqrt{2}}\varphi_r\iota_r = \frac{i}{4}(-e_1 + e_{12} - e_{134} + j)$$

These should give us the correct energy densities. We calculate operator bra $\langle u_{gr}| = \hat{u_{gr}}$ and so forth and obtain:

$$|u_{gr}\rangle\langle u_{gr}| = \frac{1}{4}(Id - e_2 - e_{34} + e_{234}) = f_{2,2}$$
$$|d_{gr}\rangle\langle d_{gr}| = \frac{1}{4}(Id - e_2 + e_{34} - e_{234}) = f_{2,3} \qquad (39)$$
$$|s_{gr}\rangle\langle s_{gr}| = \frac{1}{4}(Id + e_2 - e_{34} - e_{234}) = f_{2,4}$$
$$|\nu_{e^-}\rangle\langle \nu_{e^-}| = \frac{1}{4}(Id + e_2 + e_{34} + e_{234}) = f_{2,1}$$

blue area extender torsion momentum symbolic operator ket

blue u, d, s

$$\varphi_{bl} = \frac{1}{2}(e_3 + e_{14}) \quad \iota_{bl} = \frac{1}{\sqrt{2}}(-e_2 + e_{23})$$
$$|u_{bl}\rangle = \frac{i}{\sqrt{2}}\varphi_{bl}\hat{\iota}_{bl} = \frac{i}{4}(-e_2 - e_{23} - e_{124} + j) \qquad (40)$$
$$|d_{bl}\rangle = \frac{i}{\sqrt{2}}\hat{\varphi}_{bl}\hat{\iota}_{bl} = \frac{i}{4}(e_2 + e_{23} - e_{124} + j)$$
$$|s_{bl}\rangle = \frac{i}{\sqrt{2}}\hat{\varphi}_r\iota_r = \frac{i}{4}(e_2 - e_{23} + e_{124} + j)$$

Fix point of stabilizer $SU_{FC}(3, \mathbb{C})_{blue}$ is

$$|\nu_{e^-}\rangle = \frac{i}{\sqrt{2}}\varphi_{bl}\iota_{bl} = \frac{i}{4}(-e_2 + e_{23} + e_{124} + j)$$

These should give us the correct energy densities. We calculate operator bra $\langle u_{bl}| = \hat{u_{bl}}$ and so forth and obtain:

$$|u_{bl}\rangle\langle u_{bl}| = \frac{1}{4}(Id - e_3 - e_{14} - e_{134}) = f_{3,2}$$
$$|d_{bl}\rangle\langle d_{bl}| = \frac{1}{4}(Id - e_3 + e_{14} + e_{134}) = f_{3,3}$$
$$|s_{bl}\rangle\langle s_{bl}| = \frac{1}{4}(Id + e_3 - e_{14} + e_{134}) = f_{3,4}$$
$$|\nu_{e^-}\rangle\langle \nu_{e^-}| = \frac{1}{4}(Id + e_3 + e_{14} - e_{134}) = f_{3,1} \qquad (41)$$

Anti-fermions:

$$ch_1\ |\bar{u}_{\bar{r}}\rangle = \frac{1}{4}(e_3 - e_{13} + e_{234} - j)$$
$$|\bar{d}_{\bar{r}}\rangle = \frac{1}{4}(-e_3 + e_{13} + e_{234} - j) \quad (42)$$
$$|\bar{s}_{\bar{r}}\rangle = \frac{1}{4}(-e_3 - e_{13} - e_{234} - j)$$
$$|\bar{\nu}_{e^+}\rangle = \frac{1}{4}(-e_2 + e_{23} + e_{124} + j)$$
$$ch_3\ |\bar{u}_{\overline{gr}}\rangle = \frac{1}{4}(-e_1 - e_{12} - e_{134} + j)$$
$$|\bar{d}_{\overline{gr}}\rangle = \frac{1}{4}(e_1 + e_{12} + e_{134} + j) \quad (43)$$
$$|\bar{s}_{\overline{gr}}\rangle = \frac{1}{4}(e_1 - e_{12} - e_{134} + j)$$
$$|\bar{\nu}_{e^+}\rangle = \frac{1}{4}(-e_1 + e_{12} - e_{134} + j)$$
$$ch_5\ |\bar{u}_{\overline{bl}}\rangle = \frac{1}{4}(-e_2 - e_{23} - e_{124} + j)$$
$$|\bar{d}_{\overline{bl}}\rangle = \frac{1}{4}(e_2 + e_{23} - e_{124} + j) \quad (44)$$
$$|\bar{s}_{\overline{bl}}\rangle = \frac{1}{4}(e_2 - e_{23} + e_{124} + j)$$
$$|\nu_{e^-}\rangle = \frac{i}{4}(-e_2 + e_{23} + e_{124} + j)$$

Carrying out analogous constructions for spaces:

$$ch_2\ \ldots\ \{Id, e_1, e_{34}, e_{134}\} \quad (45)$$
$$ch_4\ \ldots\ \{Id, e_2, e_{14}, e_{124}\}$$
$$ch_6\ \ldots\ \{Id, e_3, e_{24}, e_{234}\}$$

From this we obtain the representation spaces for the charm, bottom, and top quarks. It is interesting to consider matrix representations of these isotropic Cartan spinors. To take any example, the green u-quark can be represented by the matrix:

$$|u_{gr}\rangle = \frac{1}{4}\begin{bmatrix} -1 & -1 & -1 & -1 \\ +1 & +1 & +1 & +1 \\ +1 & +1 & +1 & +1 \\ -1 & -1 & -1 & -1 \end{bmatrix} \tag{46}$$

This beautifully illustrates the zero-sum game of creation. We have to keep in mind that such a matrix summarizes all fermion subspaces that have 1-norms. The entries can be interpreted as 16 quanta emerging in 16 graded space elements. It is through this that creation of energy creates space-time. Consider, as a second example, the representation space of a charged pi-meson, say, the collection of space-time elements wherein a π^+ is moving. Usually, for instance in Wikipedia, such a particle is depicted as a term $|\pi^+\rangle = |u\bar{d}\rangle$ abstracting from the color. But, as a matter of fact, the motion space must respect all degrees of freedom of the fermion ensembles, namely, we have:

$$|\pi^+\rangle = 4i\left(|u_{red}\bar{d}_{\overline{gr}}\rangle + |u_{red}\bar{d}_{\overline{blue}}\rangle + |u_{gr}\bar{d}_{\overline{red}}\rangle + |u_{gr}\bar{d}_{\overline{blue}}\rangle + |u_{blue}\bar{d}_{\overline{red}}\rangle + |u_{blue}\bar{d}_{\overline{gr}}\rangle\right) =$$

$$|\pi^+\rangle = \frac{1}{2}(Id + e_{123} + e_{124} + e_{134} + e_{234} + e_{1234}) \tag{47}$$

which has a matrix

$$|\pi^+\rangle = \frac{1}{4}\begin{bmatrix} +1 & 0 & 0 & 0 \\ 0 & +1 & -1 & -1 \\ -1 & 0 & 0 & 0 \\ +1 & 0 & 0 & 0 \end{bmatrix} \tag{48}$$

And again we can see the specific zero sum game constituting the spinor-matrix.

Chapter 6

THE QUANTUM DATA OF GARGANTUA

Most of us probably know the movie *Interstellar*, whose first drafts, similar to *Contact*, go back to the filmmaker Lynda Obst and theoretical physicist Kip Thorne. Spielberg picked up their eight-page plot and, with Paramount Pictures, made a plan to produce the film in 2006. In 2007, Jonathan Nolan was hired to write the screenplay. A considerable working group was formed. With Kip Thorne as Consultant Executive Producer, and with Cooper recruited to pilot, the *Endurance* took off. The crew – Dr. Amelia Brand, Dr. Romilly, Dr. Doyle, and the robots TARS and CASE – have to save humanity by first inspecting some planets close to the black hole Gargantua and second by gaining insight into the hitherto unknown connection between relativity and quantum theory. They use a slingshot maneuver close to Gargantua to reach Edmund's planet. Slipping through the event horizon of Gargantua, they make a trans-dimensional journey into a huge tesseract, apparently constructed by future humans, inside the singularity of space-time. Using the *quantum data* sent by Cooper though the tesseract, his daughter Murphy solved the theoretical problems, enabling humanity's mass exodus and survival.

Cooper was able to transmit the *quantum data* via the ultra-weak but transdimensional interaction with his wristwatch in Murphy's room, and

Murphy was able to receive the signals and decode the message received by the second hand. Clumsily, Murphy threw all those pages of her precious transcript into the air, shouting "Eureka." So now I have to recount to you the most important notes about the quantum-relativity bridge as far as they are still legible to me. Although I do not have the experience of a slingshot maneuver close to Gargantua, nor do I have a good telemetric starting position, I do have precise information about the construction of the tesseract as you can easily see from the Majorana illustrations above. Also, I will use my experience, intuition, and remote sensing.

So, what are the lessons we have to learn about quantum information? What has relativity to do with quantum theory? We are living in a material world, and all physical matter is made of particles. All matter consists of interacting particles. The structure and forces of these interactions satisfy symmetries and associated symmetry breakings. But the degrees of freedom of moving matter, of elementary particles, atoms, molecules, etc., consist not only of the freedom to move along certain orientated paths in space-time, but also in breaking certain – sometimes all – symmetries. The most astonishing thing about these phenomena is that they seem to result from the geometric-algebraic symmetry laws generated by spaces with Lorentz metrics or so-called Lorentz manifolds. Our many numbers that we use to classify particles, their forces, and interactions follow on from properties of special relativity and mathematics. In short, this means that if we study Einstein's field equations and their solutions, such as the Kerr metric with its singularities, we still learn nothing about the relevant particle classification and their strong interaction. But if we just take a closer look at special relativity and, consequently, the Clifford algebra generated by flat Minkowski space alone, and if we combine that with our knowledge from high-energy physics, then it becomes apparent not only where our so-called "standard model" comes from with all its quantum numbers and classifications, but also why curved space-time and the strangely interacting Ricci tensors are elementary, real material properties that cannot easily be treated with a Reissner-Nordström metric in flat space or, say, a linearized d'Alembert field equation for rotating black holes. The matter is more subtle.

- The cosmos we perceive comes out of communication in a special zero-sum game – from the smallest nanoscopic to the largest astrophysical scales – that continuously creates, stabilizes, and breaks structures and rules of orientation. Mathematically, this process can best be understood by studying a suitable, comprehensive, geometric system with all its involutive automorphisms and anti-automorphisms. The fact that we can design this system as if it were an object stems from its universal realization in both inner and outer. The morphogenetic system of matter constitutes inner and outer space.
- Below the obvious appearance of space and time there is quantum motion that transcends time and dimension. There is movement away from time and motion beyond the perceivable observation space of a protocol, then its locus can turn out to be trans-dimensional and somewhat undecidable, but it need not be random or "principally random."
- There are events visible to some of us while invisible to others.
  - Some events are perceived by some of us some of the time.
  - Some events can be perceived by some of us all the time.
  - Some events can never be perceived by some of us.
  - Some events seem to be perceived by all of us some of the time.
  - Something is perceived by all of us all the time.
  - Most events can never be perceived by all of us.
  - Some things are sometimes perceived by no one.
  - Some can never be perceived by anyone.
  - Sometimes it seems our world can be perceived by a few.
  - Never can all of us perceive all events.
  - Sometimes no one can perceive some apparently important events, and so forth.

There are parallel timelines and separate events. There are parallel worlds that seem to us separated by the light cone. To most of us even antimatter belongs to some extra outer world that does not and should not belong to the real material world that is usually perceived by us. But antimatter with negative energy is just matter like any other matter. Actually, there is only matter. But it takes a peculiar mystery to let matter emerge within our living-oriented perception space, because at the basis there is some kind of no-thing, physically disclosed as a superposition of advanced and retarded waves unfolding from nothing. In the chapter *Drama* in *Nuclear Time Travel* [1], we read about human experience in some real interstellar journey to another star system. I cited from Len Kasten's wonderful book *Secret Journey to planet Serpo* and referred to Serpo release 24 and commentaries in Bill Ryan's homepage www.serpo.org. In his chapter *The Voyage,* Kasten has two sections, one titled *Antimatter Propulsion*, and the next one *Negative Matter Versus Positive Matter*. Both begin with quotations of corresponding entries of the team commander in his diary. This designation of *negative versus positive matter* hits the important point exactly.

In *Primordial Space* [49, chapter 2], I have shown how the Minkowski algebra generated by the Minkowski space as a Clifford algebra can be understood as a graded differential Lie algebra with brackets, having dimension 15 and rank 3. Now recall that there is a split in the whole mathematical theory that makes a difference between matter and antimatter. This split is supported by the partitioning:

$$Cl_{3,1} \oplus i\, Cl_{3,1} \subset SL(4, \mathbb{C}) \qquad (49)$$

where the whole algebra $Cl_{3,1}$ comes up as a real part in a complex number. That real part is a real 4×4 matrix of the Majorana algebra $Mat(4, \mathbb{R})$ and the imaginary part is given by a matrix with imaginary entries only, that is, the product $i\, Cl_{3,1}$ with the imaginary unit $\sqrt{-1}$. All our fermions are isotropic multi-vectorial Cartan spinors ξ in the imaginary part $i\, Cl_{3,1}$, so that their product with the grade-involuted $\hat{\xi}$

gives back the associated real energy density, namely the idempotent $f = \xi\hat{\xi} \in Cl_{3,1}$, such that the spinor is located in a minimal left ideal of the algebra, $\xi \in f(\mathbb{C}\otimes Cl_{3,1})$. That this possibility became *fact in theory* was based on the solution to this mathematical problem. It could be solved by the aid of the software Maple Clifford developed by Bertfried Fauser and Rafal Ablamowicz. The Cartan spinors of the associated anti-fermions are in the real part $Cl_{3,1} \simeq Mat(4, \mathbb{R})$ and their energy densities are negative idempotents again in the Minkowski algebra $Mat(4, \mathbb{R})$ over real numbers. Hence, we indeed have a decomposition of matter as …

NEGATIVE ENERGY VERSUS POSITIVE ENERGY

"633 wants to see the engines. MVC takes four of us to the engine room or whatever they wish to call the room. It contains large, very large metal containers. They are in a circle, with the ends of each pointing into the center. Many pipes or some type of large tubes connects them. In the center of these containers is a copper colored coil or something looking like a coil. There is a bright light being shined from a point above into the center of the coil. We hear a very dull hum, but no major loud sounds. 661 thinks it is a negative matter versus positive matter system" [67].

Does that not remind us of that "deeper look into the inner of the circular, disc-shaped planform" documented in General Twinings' "preliminary examination D333.5 ID from 15th July 1947," quoted in Nuclear Time Travel [1, p. 9]? The description of the engine room is as follows: "A doughnut shaped tube approximately thirty-five feet in diameter, made of what appears to be a plastic material, surrounding a central core … A large rod centered inside the tube, was wrapped in a coil of what appears to be of copper material, run through the circumference of the tube. This may be the reactor control mechanism or a storage battery. There were no moving parts discernible within the power room." Ebens' spacecrafts use antimatter reactors, current sources and batteries and small, low-energy, anti-particle accelerators.

Apart from many forms of light, real physical matter seems to get by with two quarks, u and d, and their anti-quarks to make protons and neutrons, we need plenty of electrons, positrons, some genus of π-mesons, muons and (anti)electron-neutrinos. Leaving the comparatively simple world of everyday matter, we enter the particle zoo of HEPhy with its huge lists of flavorless, flavored, exotic, pseudoscalar, vector mesons and neutral kaons. There is no unity in HEPhy as it restricts orientation to C-T-P symmetry and C-T-P violation. But there are thousands of more (anti)automorphisms in physics and subnuclear resonances need not be high-energy to become relevant. There are giant nuclear multi and quadrupole resonances at astonishingly low energies. Such resonances within the nuclei excite the torsion component in the multivectorial quark spinor. That is, a u-quark can be excited to much higher spin components than those described by isospin 1. The space-time spinors presented here as *Quantum Data of Gargantua* allow for:

- A rather unified theory of elementary material phenomena,
- without lists of flavorless, flavored, exotic, pseudoscalar, vector mesons and similar superfluous differences. Anyhow, there are mesons and neutrinos having more or less mass.
- Extension and torsion can be excited in advanced and retarded fermion waves. They come out of nothing and come in from nowhere and superimpose, creating energy density.
- Subnuclear fermion resonances cause a warp and a time wave.

Consider Cartan spinor of a red u-quark:

$$|u_r\rangle = \frac{i}{\sqrt{2}}\varphi_r \hat{\iota_r} = \frac{i}{4}(e_3 - e_{13} + e_{234} - j)$$

$$\langle u_r| = -\frac{i}{4}(e_3 + e_{13} + e_{234} + e_{1234}) \qquad (50)$$

with product

$$|u_r\rangle\langle u_r| = \frac{1}{4}(Id - e_1 - e_{24} + e_{124}) = f_{1,2}$$

having matrix representations

$$|u_r\rangle = \begin{bmatrix} 0 & 0 & 0 & 0 \\ 0 & 0 & 0 & -\mathrm{i} \\ 0 & 0 & 0 & 0 \\ 0 & 0 & 0 & 0 \end{bmatrix} \langle u_r| = \begin{bmatrix} 0 & 0 & 0 & 0 \\ 0 & 0 & 0 & 0 \\ 0 & 0 & 0 & 0 \\ 0 & \mathrm{i} & 0 & 0 \end{bmatrix} \tag{51}$$

hence

$$|u_r\rangle\langle u_r| = \begin{bmatrix} 0 & 0 & 0 & 0 \\ 0 & 1 & 0 & 0 \\ 0 & 0 & 0 & 0 \\ 0 & 0 & 0 & 0 \end{bmatrix} \cdots f_{1,2}$$

The isotropic spinor is elementary in a mathematical sense and physically as it represents the elementary, allegedly extensionless, particles of energy. Mathematically, it is minimal as an element in a minimal ideal having minimal dimension in a hypercomplex space with dimension sixteen. These elementary containers of energy provide the minimal representation spaces of spinors in the sense proposed by Elié Cartan, Marcel Riesz, Claude Chevalley, and Pertti Lounesto. They just give us basic combinations of space elements having different grades necessary to construct fermions with all the desired quantum numbers, while compatible with demands of special and general relativity. We must differ between those elementary subspaces of representation and the related function spaces of hypercomplex variables from minimal ideals in high-dimensional hypercomplex spaces such as $Cl_{3,1} \oplus i\, Cl_{3,1} \subset SL(4,\mathbb{C})$. In German, "wir begeben uns hier in das Fachgebiet der *Bereichstheorie*" (English: here, we enter the field of *domain theory*). It would probably take more than one book to describe this area adequately. We are dealing here with completely new approaches. But, for now, we can at least try to show the new physical situation we are in, and we do so with the means of minimal representation spaces.

The isotropic spinor is a possibility, its energy a reality. It annihilates itself, but creates an energy inflow within the first color space ch_1 when superimposed by the grade involuted wave. To make the energy of the spinor real, superposition with the retarded wave is one possibility, and there is a second one that is based on a connection with negative energy that, in civil engineering, we have not yet come upon since it is based on the excitation of isospin resonances in heavy symmetric nuclei. For a long time we have marveled at the great frequency and enormous amount of antimatter used in the deployment of extraterrestrial devices. But, as time went on, it became clearer to us how things work. Within the whole network of intelligent awareness and cosmic energy, there exist wonderful and surprising modes of transmutation. To illustrate this, let us try to quantize the excitation of isospin in a fermion in a way that seems somewhat familiar to us. Consider spinor components:

$$\varphi_{red} = \frac{1}{2}(e_1 + e_{24}),\ \iota_{red} = \frac{1}{\sqrt{2}}(e_3 + e_{13}) \tag{52}$$

and rename them in the style of angular momentum algebra:

$$J_3 = \varphi_{red} = \frac{1}{2}(e_1 + e_{24}) \quad J_+ = \hbar\,\iota_{red} = \frac{\hbar}{\sqrt{2}}(e_3 + e_{13})$$
$$J_- = \frac{\hbar}{\sqrt{2}}(e_3 - e_{13}); \tag{53}$$

Then, we actually verify the well-known commutation relations:

$$[J_3, J_+] = J_+ \quad [J_3, J_-] = -J_- \tag{54}$$

from which we can derive the first and second components of the angular momentum:

$$J_1 = \frac{1}{2}(J_+ + J_-) = \frac{\hbar}{\sqrt{2}} e_3 \text{ and}$$
$$J_2 = -\frac{i}{\sqrt{2}}(J_+ - J_-) = -\frac{i\,\hbar}{\sqrt{2}} e_{13}. \tag{55}$$

Together with $J_3 = \frac{1}{2}(e_1 + e_{24})$ these allow us to define the total angular momentum multivector

$$J^2 \stackrel{\text{def}}{=} J_1^2 + J_2^2 + J_3^2 \text{ that turns out to be:}$$
$$J^2 = \hbar^2 Id + \frac{1}{2}(Id + e_{123}), \tag{56}$$

and this actually commutes with the third component as we have:

$$[J^2, J_3] = 0$$

But it does not commute with the first and second component as should traditionally be expected from matrices of angular momentum, namely we get:

$$[J^2, J_1] = -\frac{\hbar}{\sqrt{2}} e_{1234} \qquad [J^2, J_2] = \frac{i\,\hbar}{\sqrt{2}} e_{234} \qquad \text{together with} \tag{57}$$
$$[J^2, J_+] = -\frac{\hbar}{\sqrt{2}}(e_{234} + e_{1234})$$
$$[J^2, J_-] = \frac{\hbar}{\sqrt{2}}(e_{234} - e_{1234})$$

so that the ladder operators of isospin seem additionally to excite a creation of space-time 4-volume. What is going on here? In the original formalism of orbital, spin, and isospin angular momentum, we are spoiled with pleasant commutation relations, namely J^2 commuted with all of them, with J_1, J_2, J_3, and the ladder operators. We were free to select any of the three, but by convention we chose the third to construct an eigenbasis for the pair J^2, J_3 having eigenvalues, say, $|j, m\rangle$. This is somewhat difficult now because the selection of polarized spinors in minimal ideals is arbitrary and does not equip us with appropriate hypercomplex function spaces. Yet we can show the new events that enter the stage. In our case of isotropic space-time spinors, it turns out, from the beginning, that the extensor $J_3 = \frac{1}{2}(e_1 + e_{24})$ defines a privileged orientation of motion. The ladder operators raise or lower what we had decided to call the *magnetic quantum number*, that is, the *m* in the $|j, m\rangle$. There was a famous equation of normalization:

$$J_{\pm}|j, m\rangle = c_{\pm}|j, m \pm 1\rangle \quad (58)$$

This helped us to calculate Clebsch-Gordan coefficients. It could be solved by the aid of a no-less famous equation that read, in its original form (together with the reduced Planck constant):

$$J^2 = J_+J_- + J_3^2 - \hbar\, J_3 \quad (59)$$

which gave us the familiar result of $c_{\pm} = \hbar\sqrt{j(j+1) - m(m \pm 1)}$. This undergoes considerable transmutations. Since J^2 now does not commute with J_1, J_2, another equation gains relevance. This reads:

$J^2 = \frac{1}{2}\{J_+, J_-\} + J_3^2$ with the anti-commutator brackets $\{..\}$, namely in detail

$$J^2 = \frac{1}{2}(J_+J_- + J_-J_+) + J_3^2 \quad (60)$$

It seems the algebra provides a simple central extension of the Lie algebra $su(3, \mathbb{C})$ that transforms an u_{red} in the red corner of the hexagram. Why so? Well, we have:

$$J_+J_- = \hbar^2(Id + e_1) \;\; J_-J_+ = \hbar^2(Id - e_1) \quad (61)$$

Hence the commutator is $[J_+, J_-] = 2\hbar^2\, e_1$ and the anti-commutator is a multiple of the algebra's identity $\{J_+, J_-\} = 2\hbar^2\, Id$. It is therefore that all isotropic Cartan spinors that we have designed so far to represent elementary particles, namely, quarks, electrons, mesons and neutrinos are equipped with isospin-shift-operators whose anti-commutators commute with the whole Clifford algebra. Here mathematics shows us its sense for dialectics. While physicists prove certain elementary particles have no extension, mathematics shows us central extensions of their associated small Lie algebras that expand over the whole Minkowski algebra.

Beginning with what we have, how could we understand excitation of the isospin torsion momentum of a fermion? We found out that J^2 and J_3 commute. So we could consider the possibility of a common eigenbasis in hypercomplex function space. Further we have that equation that would lead us to some normalization matrix, namely

$J^2 - J_3^2 = \frac{1}{2}(J_+J_- + J_-J_+)$ which together with commutator $[J_+, J_-] = 2\hbar^2\, e_1$ gives us the most important entry to the problem

$$J_-J_+ = J^2 - J_3^2 - \hbar^2 e_1 \tag{62}$$

Recall that e_1 defines one of the main orientations of the extension, and we denoted that direction by J_3 to accentuate the analogy with traditional angular momentum. However, the extension represents a preferred direction of motion which brings on a natural coupling between nuclear isospin and direction of travel. Considering normalization equation $J_+|j,m\rangle = C_+|j,m+1\rangle$ with a normalization-matrix that depends on extension and torsion, we obtain

$$\langle j,m|\, J_+^{\dagger}J_+ \big|j,m\rangle = \langle j,m|\, J_-J_+|j,m\rangle = \langle j,m\big|J^2 - J_3^2 - \hbar^2 e_1\big|j,m\rangle = \tag{63}$$

$= \hbar^2(j(j+1) - m^2 - e_1)\langle j,m|j,m\rangle$ from which $C_+^2 = \hbar^2\big((j+m)(j-m) + j\big)Id - \hbar^2 e_1$

This can abstractly be solved for C_+ because C_+^2 having form $(a\, Id - b\, e_1)$ is almost an idempotent. Hence C_+ has that idempotent form too, say,$(x_1\, Id - x_2\, e_1)$ and can be found by coefficient comparison. As we carry out the same rigor for $J_-|j,m\rangle = C_-|j,m-1\rangle$, we find that a transition from raising to lowering torsion momentum is bound to a flip of direction:

$$C_\pm^2 = \hbar^2\big((j+m)(j-m)+j\big)Id \mp \hbar^2 e_1 \tag{64}$$

Flipping torsion shift operators correlates with a reversion of spatial line element $\hbar^2 e_1$. If we recall *Nuclear Time Travel* [1, p. 16] where it was reported that "a single, central, spherical shaft or pole running through the thickness of the craft and connecting the upper and lower outer hulls which also divided into equal thirds. The pole running through the saucer was about forty-one (41) feet long and about sixteen (16) inches in diameter. It was a perfect cylinder of a solid carbon/zinc alloy. When maintained or stored in a vertical position this pole has no special power (of electricity). But when laid down at the Hart Canyon site in its side, the pole began to drain electrical current from machines, generators and batteries and to broadcast a powerful (magnetic ...) which damaged sensitive instruments and magnetized tools. For ... had to be transported in an upright fashion." Here we are observing isospin coherence locked with direction, and there is more to that mystery. Interestingly, the anti-commutation relation $\{J_+, J_-\} = 2\hbar^2\, Id$ that extends over the whole algebra leads to:

$$\langle j,m|J^2 - J_3^2 - \hbar^2 e_1|j,m\rangle = \hbar^2(j(j+1) - m^2 - 2Id)\langle j,m|j,m\rangle = 0 \tag{65}$$

And to the quadratic equation:

$$j^2 - j + (2 - m^2) = 0 \tag{66}$$

For low excitation states (m=2) we get two solutions for the quantum number j, namely $j_a = -1$ and $j_b = +2$; with quantum-numbers $m = 2, j_b = +2$ indicating gravitational quadrupole radiation. I agree these are theoretical considerations. But I imagine that such oscillations of energy in graded minimal subspaces are real and once brought into coherent form of giant resonances, these can be stabilized and locked with forward motion of the extension-space element. Coherent quadrupole resonances of isospin will generate a strong gravitational wave on which the enlarged closed space-time containments of the spacecraft can ride. With the spin 2 graviton we are back to Halpern 1968. However, he got this result by linearizing Einstein's field equations in flat space. In our case, the existence of the graviton and its associated quadrupole field emerges from the excitation of subnuclear isospin, in particular from control over the torsion momenta of partons.

A Final Note on This Type of Literature

You saw and you read, I let children have their say, and esoteric people who are often called crazy. I agreed with scientists who are at the very edge of society. I cared for the marginal minds. But if you think that by doing so I have endangered myself, you are wrong. The things I have described to you are correct. I was depressed, and sometimes even horrified to see how these extremely competent, talented, and creative people were suspected and marginalized. That strengthened me all the more and motivated me to strengthen their back. It made me particularly weak when I saw that even Linda Moulton Howe [68] started posting suspicious and hostile contributions against Burisch. The Earthfiles are perhaps the most comprehensive and most important documents about alien presence. Finally, Linda Moulton Howe has at least supplemented and relativized the document in question, writing *"Neither A Negative Nor Positive Yet Proved.* So far, the July 16, 2002 Chapter 7 bankruptcy does not prove the negative that Dan Burisch has lied about doing microbiology work for the MJ government group. But there is no hard proof for the positive either that Burisch has worked for the government. When I asked Bill Hamilton about the bankruptcy issue, he said ..."

My positive attitude towards communication using words is exactly what Birgit Birnbacher has said about literature: "I think that in literature

– … I am not at all interested in the question of whether it is invented or how true something or how untrue it is, because ultimately it is always about a kind of truthfulness, and I always find it really moving when literature opens such a moment that neither the writer nor the reader has experienced, but both of them know very well that it is just that, that it feels exactly the same. This is actually a form of truthfulness that I have always liked very much."[25]

[25] In a radio interview on July 28, 2020: "Ich habe mich immer schon für Außenseiter stark gemacht" – Writer and sociologist Birgit Birnbacher on the willingness to empathize with others and the truthfulness of literature.

APPENDIX: SIX FUNDAMENTAL FERMION STABILIZER GROUPS OF CLIFFORD ALGEBRA OF MINKOWSKI SPACE

The table has eight rows for the generators of the special linear group $SL(3,\mathbb{R})$ – the "Heisenberg group" – and six columns for the color spaces $ch_1, \ldots, ch_6$. Multiplying rows 2, 5, and 7 with the imaginary unit, we apply the so-called Weyl trick and carry $SL(3,\mathbb{R})$ to the special unitary group $SU(3,\mathbb{C})$. The latter was introduced by Murray Gell-Mann to classify elementary particles in terms of quark multiplets. Algebraic hypercharge in $su(3,\mathbb{C})$ as given by row number nine, the λ_8^*, arises from a linear combination of λ_3 and λ_8. [1. p. 107].

Appendix

	ch_1	ch_2	ch_3	ch_4	ch_5	ch_6
λ_1	$\frac{-e_{34}+e_{134}}{2}$	$\frac{-e_{24}+e_{124}}{2}$	$\frac{-e_{14}-e_{124}}{2}$	$\frac{-e_{34}+e_{234}}{2}$	$\frac{-e_{24}-e_{234}}{2}$	$\frac{-e_{14}-e_{134}}{2}$
$\lambda_2(\times i)$	$\frac{-e_{23}+e_{123}}{2}$	$\frac{e_{23}-e_{123}}{2}$	$\frac{e_{13}+e_{123}}{2}$	$\frac{-e_{13}-e_{123}}{2}$	$\frac{-e_{12}+e_{123}}{2}$	$\frac{e_{12}-e_{123}}{2}$
λ_3	$\frac{-e_{24}+e_{124}}{2}$	$\frac{-e_{34}+e_{134}}{2}$	$\frac{-e_{34}+e_{234}}{2}$	$\frac{-e_{14}-e_{124}}{2}$	$\frac{-e_{14}-e_{134}}{2}$	$\frac{-e_{24}-e_{234}}{2}$
λ_4	$\frac{-e_3-e_{234}}{2}$	$\frac{-e_2+e_{234}}{2}$	$\frac{-e_1+e_{134}}{2}$	$\frac{-e_3-e_{134}}{2}$	$\frac{-e_2-e_{124}}{2}$	$\frac{-e_1+e_{124}}{2}$
$\lambda_5(\times i)$	$\frac{-e_{13}-J}{2}$	$\frac{-e_{12}+J}{2}$	$\frac{e_{12}-J}{2}$	$\frac{-e_{23}+J}{2}$	$\frac{-e_{23}-J}{2}$	$\frac{e_{13}+J}{2}$
λ_6	$\frac{e_2+e_{14}}{2}$	$\frac{e_3+e_{14}}{2}$	$\frac{e_3+e_{24}}{2}$	$\frac{e_1+e_{24}}{2}$	$\frac{e_1+e_{34}}{2}$	$\frac{e_2+e_{34}}{2}$
$\lambda_7(\times i)$	$\frac{e_4+e_{12}}{2}$	$\frac{+e_4+e_{13}}{2}$	$\frac{e_4+e_{23}}{2}$	$\frac{e_4-e_{12}}{2}$	$\frac{e_4-e_{13}}{2}$	$\frac{e_4-e_{23}}{2}$
λ_8	$\frac{-e_1+e_{24}}{2}$	$\frac{-e_1+e_{34}}{2}$	$\frac{-e_2+e_{34}}{2}$	$\frac{-e_2+e_{14}}{2}$	$\frac{-e_3+e_{14}}{2}$	$\frac{-e_3+e_{24}}{2}$
λ_8^*	$\frac{-2e_1+e_{24}+e_{124}}{2\sqrt{3}}$	$\frac{-2e_1+e_{34}+e_{134}}{2\sqrt{3}}$	$\frac{-2e_2+e_{34}+e_{234}}{2\sqrt{3}}$	$\frac{-2e_2+e_{14}-e_{124}}{2\sqrt{3}}$	$\frac{-2e_3+e_{14}-e_{134}}{2\sqrt{3}}$	$\frac{-2e_3+e_{24}-e_{234}}{2\sqrt{3}}$

REFERENCES

[1] Schmeikal, B. (2019). *Nuclear Time Travel and the Alien Mind.* Nova Science Publishers: New York.

[2] Schmeikal, B. (2018). *Einstein's Blunder.* https://www.researchgate.net/profile/Bernd_Schmeikal.

[3] Angelucci, O. (1955). *The Secret of the Saucers.* http://www.exopoliticshongkong.com/uploads/secretofthesaucers_by_Orpheo_Angelucci_1955.pdfhttps://www.kobo.com/ww/en/ebook/the-secret-of-the-saucers

[4] Lounesto, P. (2001). *Clifford algebras and Spinors*; Cambridge University Press: Cambridge.

[5] Schmeikal, B. (2014). *Decay of Motion - The Anti-Physics of Space-Time.* Nova Science Publishers: New York.

[6] Schmeikal, B. (2016). Four Forms Make a Universe, *Adv. Appl. Cliff Alg,* 26 (3), pp 889-911. DOI 10.1007/s00006-015-0551-z, http://link.springer.com/article/10.1007/s00006-015-0551-z.

[7] Schmeikal, B. (2014). Tessarinen, Nektarinen und andere Vierheiten, Beweis einer Beobachtung von Gerhard Opfer. *Mitt Math Ges Hamburg*, 34, pp 1–28.

[8] Wiesengrün, M. (2018). *Mein UFO-Erlebnis auf Rügen.* [*My UFO Experience on Rügen.*] Ventla: Gütersloh.

[9] Fomin, V. M. (2018). *Physics of Quantum Rings*. Springer International: New York.

[10] Pareto principle, *Wikipedia, The Free Encyclopedia,* https://en.wikipedia.org/w/index.php?title=Pareto_principle&oldid=954412306 (accessed May 17, 2020).

[11] Pareto distribution, Wikipedia, *The Free Encyclopedia*, https://en.wikipedia.org/w/index.php?title=Pareto_distribution&oldid=954036990/ (accessed May 17, 2020).

[12] Employment in medium-sized enterprise. *Statista*, https://de.statista.com/statistik/daten/studie/261433/umfrage/mittelstaendische-unternehmen-in-deutschland-nach-anzahl-der-beschaeftigten/ (accessed May 17, 2020).

[13] Kajfež-Bogataj, L., Müller K. H., Svetlik, I., Toš, N. eds. (2010). *Modern RISC-Societies. Towards a New Paradigm for Societal Evolution*. edition echoraum: Wien.

[14] Macy conferences, Ninth Cybernetics Conference, 20–21 March 1952. *Wikipedia, The Free Encyclopedia,* https://en.wikipedia.org/w/index.php?title=Macy_conferences&oldid=928765814 (accessed May 15, 2020).

[15] Liste der größten Unternehmen der Welt. [List of largest NTTenterprizes worldwide]. *Wikipedia, Die freie Enzyklopädie*. Editing status: May 4, 2020, 00:26 UTC. URL: https://de.wikipedia.org/w/index.php?title=Liste_der_gr%C3%B6%C3%9Ften_Unternehmen_der_Welt&oldid=199592751 (accessed May 18, 2020, 09:58 UTC).

[16] List of largest companies by revenue. *Wikipedia, The Free Encyclopedia,* https://en.wikipedia.org/w/index.php?title=List_of_largest_companies_by_revenue&oldid=957231261 (accessed May 18, 2020).

[17] Rabe, L. (2020). Umsatz von Alphabet weltweit bis 2019, [Revenue of Alphabet worldwide till 2019]. 06.02.2020. *Stastita*. https://de.statista.com/statistik/daten/studie/74364/umfrage/umsatz-von-google-seit-2002/.

[18] "How the World's Biggest Companies Fight to Stay Ahead." *Fortune*. Archived from the original on July 22, 2019. (retrieved July 22, 2019).

[19] Forbes Global 2000. *Wikipedia.* https://en.wikipedia.org/wiki/Forbes_Global_2000

[20] Clement, J. Google: annual operating income 2013-2019. *Statista.* https://www.statista.com/statistics/513129/operating-income-google/ (accessed Feb. 5, 2020).

[21] Coleman, J. S. (1973). *The mathematics of collective action.* Aldine: Chicago.

[22] Coleman, J. S. (1990). *Foundations of Social Theory.* Belknap Press of Harvard University Press: Cambridge, MA.

[23] Scenarios for the Future of Technology and International Development, report by *The Rockefeller Foundation and Global Business Network.* May 2010.

[24] Silberjunge, *"Silberjunge" Coronakrise & schockierendes Rockefeller-Papier!* [*"Silver Boy" Corona Crisis & Shocking Rockefeller Paper!*] Elite (Bill Gates, George Soros ...) entlarvt sich [Elite expose itself]! https://youtu.be/Lw_olzJAnj0 (accessed April 4, 2020).

[25] Johns Hopkins Bloomberg School of Public Health. *Wikipedia, The Free Encyclopedia,* https://en.wikipedia.org/w/index.php?title=Johns_Hopkins_Bloomberg_School_of_Public_Health&oldid=947231394 (accessed April 4, 2020).

[26] *Development and the Information Age: Four Global Scenarios for the Future of Information and Communication Technology (1997).* International Development Research Centre (Canada), United Nations Commission on Science and Technology for Development IDRC. https://books.google.at/books/about/Development_and_the_Information_Age.html?id=7-WKAnPExxgC&redir_esc=y.

[27] Dorey, E. (2020). *Dossiers coronavirus: suivez les recherches sur l'épidémie.* [*Coronavirus file: follow the epidemic investigation*]. https://www.science-et-vie.com/archives/la-pandemie-mondiale-36315 (accessed April 4, 2020).

[28] Simmank, J. (2020). *Der heimliche WHO-Chef heißt Bill Gates.* [*The secret WHO chief is called Bill Gates*] https://www.zeit.de/wissen/gesundheit/2017-03/who-unabhaengigkeit-bill-gates-film (accessed April 4, 2020).

[29] *Letters to the Editor*, Some Comments on the Coronavirus Pandemic Introduction by the Editors, January 2020, DOI: 10.13140/RG.2.2.26466.76487, Project: Creative Intelligence and Human Progress, Lab: Bernd Schmeikal's Lab.

[30] Chomsky, N. *Coronavirus - What is at stake?* | DiEM25 TV. https://www.youtube.com/watch?v=t-N3In2rLI4&feature=youtu.be https://www.youtube.com/watch?v=t-N3In2rLI4 (Noam Chomsky)

[31] Abel, T., McQueen, D. (2020). COVID-19 pandemic calls for spatial distancing and social closeness: not for social distancing! The *International Journal of Public Health, 65, p.231.* https://doi.org/10.1007/s00038-020-01366-7.

[32] Revolutionäre Demokratische Front der Äthiopischen Völker. [Revolutionary Democratic Front of the Ethiopian Peoples.] *Wikipedia, Die freie Enzyklopädie*. Editing status: 3. Februar 2020, 23:54 UTC. URL: https://de.wikipedia.org/w/index.php?title=Revolution%C3%A4re_Demokratische_Front_der_%C3%84thiopischen_V%C3%B6lker&oldid=196476912/ (accessed May 5, 2020, 07:25 UTC)

[33] Isaacson, J. D. (1981). Autonomic string-manipulation system, US Patent No. 4286330, August 25, 1981. https://patents.google.com/patent/US4286330A/en (accessed September 2, 2020). (This was a continuation of application Ser. No. 674,658, filed April 7, 1976 with no cross-references to related applications, and a single relevant reference to a related disclosure document entitled Autonomic String-Manipulation System, No. 045773, filed on December 29, 1975).

[34] Isaacson, J. D., Kaufmann, L. H. (2016). Recursive Distinctioning. International Space Development Conference - 2016 in San Juan Puerto Rico on May 22, 2016. *Journal of Space Philosophy 5, no.*

1, pp. 9 - 65. http://homepages.math.uic.edu/~kauffman/RD.html http://www.cybsoc.org/index.html.

[35] Isaacson, J. D. (1987). *Dialectical Machine Vision, Applications of Dialectical Signal-Processing to Multiple Sensor Technologies.* Report prepared for the Strategic Defense Initiative Organization Office of Naval Research, Arlington, VA, 35 p.

[36] Krishnamurti, J. (1976). *Krishnamurti's notebook.* Foreword by Lutyens, M. (1st UK ed.). Gollancz: London.

[37] Krishnamurti, J. (1998). *Über Leben und Sterben* [*About life and death*]. Fischer: Frankfurt am Main.

[38] *Documentary USA 2016: A thriller in contemporary history!* arte media library. Damascus, USA. .https://programm.ard.de/?sendung=287243226705276.

[39] Strauss, W., Howe, N. (2009). *The Fourth Turning: What the Cycles of History Tell Us About America's Next Rendezvous with Destiny*. Broadway Books: New York.

[40] Fremont-Smith, F. (1961). The Interdisciplinary Conference. *AIBS Bulletin, 11, 2, pp. 17-20.* https://doi.org/10.2307/1292675 .

[41] Information explosion. *Wikipedia, The Free Encyclopedia.* https://en.wikipedia.org/w/index.php?title=Information_explosion &oldid=941227863 (accessed July 22, 2020).

[42] Wayback Machine. *Wikipedia, The Free Encyclopedia*. https://en.wikipedia.org/w/index.php?title=Wayback_Machine&oldid=968114820 (accessed July 22, 2020).

[43] *Oh-that's-what-my-dad-built, a statement by Will Uhouse, and Kerry Lynn Cassidy interviewing Dan Burisch*. https://rosettasister.wordpress.com/2011/01/16/oh-thats-what-my-dad-built-will-uhouse-aka-sheepy-sheephogan/

[44] Burisch, D. (2007). *Stargate Secrets – Part 1, Stargate Secrets : Dan Burisch revisited*, *A video interview with Dan Burisch*, Las Vegas, June 2007. Shot, edited and directed by Kerry Lynn Cassidy Interview transcript. http://www.projectcamelot.org/lang/en/dan_burisch_stargate_secrets_interview_transcript_1_en.html

[45] Burisch, D. (2015). Stargates - *Through the Looking Glass*. https://www.thelivingmoon.com/42stargate/03files/Project_Looking_Glass_LANL.html.

[46] Feldman, B. J., Bigio, I. J., Fisher, R. A., Phipps, C.R. (Jr.), Watkins, D. E. and Thomas, S. J. (1982). *Through the Looking Glass with Phase Conjugation*, Los Alamos Science/Fall 1982, Pegasus PDF Database, https://permalink.lanl.gov/object/tr?what=info:lanl-repo/lareport/LA-UR-82-5213.

[47] KyeoReh, L., Junsung, L., Jung-Hoon, P., Ji-Ho, P., YongKeun, P. (2015). One-Wave Optical Phase Conjugation Mirror by Actively Coupling Arbitrary Light Fields into a Single-Mode Reflector, *Physical Review Letters*, 115(15). https://www.researchgate.net/publication/276296190_One-Wave_Optical_Phase_Conjugation_Mirror_by_Actively_Coupling_Arbitrary_Light_Fields_into_a_Single-Mode_Reflector (accessed July 25, 2020).

[48] Schmeikal, B. (2011). The Universe of Spacetime Spinors. *Proceedings of the 9th International Conference on Clifford Algebras and their Applications in Mathematical Physics*. Digital Proceedings (ICCAA9). Bauhaus University, Weimar.

[49] Schmeikal, B. (2012). *Primordial Space – Pointfree Space and Logic Case*. Nova Science Publishers: New York.

[50] Halpern, L. and Laurent, B. (1964). On the gravitational radiation of microscopic systems. *Nuovo Cimento* 33, pp. 728-751.

[51] Baker, R. M. L., Jr. (2000). Preliminary Tests of Fundamental Concepts Associated with Gravitational-Wave Spacecraft Propulsion, *American Institute of Aeronautics and Astronautics, Space 2000 conference and exposition, Paper Number 2000-5250*, September 20, 2000.

[52] Fontana, G. and Baker, R. M. L., Jr. (2003). The High-Temperature Superconductor (HTSC) Gravitational Laser (GASER). *Proceedings of the Gravitational-Wave Conference*, edited by P. Murad and R. Baker, The MITRE Corporation, Mclean, Virginia, May 6-9, Paper HFGW-03-107.

[53] Alekseev, G. A. and Griffiths, J. B. (1996). Exact solutions for gravitational waves with cylindrical, spherical and toroidal wavefronts, *Classical and Quantum Gravity* 13, pp. 2191-2209.

[54] Fontana, G. (2012). *High temperature superconductors as quantum sources of gravitational waves: The HTSC gaser*, January 2012, doi:10.2174/97816080539951120101 0058, https://www.researchgate.net/publication/286619454_High_temperature_superconductors_as_quantum_sources_of_gravitational_waves_The_HTSC_gaser.

[55] Baker, R. M. L., Jr. and Baker, B. S. (2012). *Physics Procedia* 38, pp. 288 – 297. The Space Technology & Applications International Forum (STAIFII) Gravitational wave generator apparatus.

[56] Einstein, A. (1918). *Sitzungsberichte d. preuss.* Akad.

[57] Einstein, A. (1916). *Sitzungsberichte d. preuss.* Akad.

[58] Fontana, G. (2004). *Design of a Quantum Source of High-Frequency Gravitational Waves (HFGW) and Test Methodology.* arXiv:physics/0410022v1 [physics.gen-ph], https://www.researchgate.net/publication/2171404_Design_of_a_Quantum_Source_of_High-Frequency_Gravitational_Waves_HFGW_and_Test_Methodology.

[59] Schmeikal, B. (1996). The generative process of space-time and strong interaction - quantum numbers of orientation. Ablamowicz, R., Lounesto, P. Parra, J. M. (eds.), *Clifford Algebras with Numeric and Symbolic Computations,* pp. 83-100. Birkhäuser: Boston.

[60] Schmeikal, B. (2001). Minimal Spin Gauge Theory -Clifford Algebra and Quantumchromodynamics. *Advances in Applied Clifford Algebras* 11(1), pp. 63-80. Follow journal. Project: Juwels in Applied Clifford Algebra. https://www.researchgate.net/publication/242119234_MINIMAL_SPIN_GAUGE_THEORY_-Clifford_Algebra_and_Quantumchromodynamics.

[61] Schmeikal, B. (2004). Transposition in Clifford Algebra; In *Clifford Algebras – Applications to Mathematics Physics and Engineering;* Ablamowicz, R., ed.; Birkhäuser: Boston, pp. 351-372.

[62] Schmeikal, B. (2013). On Motion. *Clifford Analysis, Clifford Algebras and their Applications.* (CACAA) 2(1), Cambridge, UK.

[63] Friebe. E. (1988). Was sind physikalische Gesetze? [What are physical laws?] *raum & zeit*, 32/88, S. 88-91, https://ekkehard-friebe.de/Gesetz-2, (accessed Feb. 17, 2021).

[64] Chandler, C. (2010). *Vacuum Conductivity.* http://qdl.scs-inc.us/2ndParty/Pages/8819.html (accessed Feb. 17, 2021).

[65] Rajagopal, K., Wilczek, F. (1986). Enforced Electrical Neutrality of the Color-Flavor Locked Phase *Physical Review Letters* 86(16):3492-5. doi: 10.1103/PhysRevLett.86.3492 Enforced Electrical Neutrality of the Color-Flavor Locked Phase | Request PDF (researchgate.net).

[66] Schmeikal, B. (2005). Algebra of Matter. *Advances in Applied Clifford Algebra* 15, No. 2. doi: 10.1007/s00006-005-0013-0. Algebra of matter | Request PDF (researchgate.net).

[67] Ryan, B. (2005). *Posting Eleven by Anonymous* (21 December, 2005). Regarding our team members. http://www.serpo.org/release11.php.

[68] Moulton Howe, L. (2020). *The Chapter 7 Bankruptcy of Dan and Deborah Burisch*, Posted on July 13, 2004 © 2020, https://www.earthfiles.com/2004/07/13/the-chapter-7-bankruptcy-of-dan-and-deborah-burisch/.

AUTHOR'S CONTACT INFORMATION

Dr. Bernd Schmeikal
Univ. Doz Univerisity of Vienna
Wien, Austria
Email: schmeika@wu.ac.at.

INDEX

A

C

D

E

F

G

U

X

Z